# OPTICAL MODELING AND SIMULATION OF
# THIN-FILM
# PHOTOVOLTAIC
# DEVICES

# OPTICAL MODELING AND SIMULATION OF THIN-FILM PHOTOVOLTAIC DEVICES

Janez Krč

Marko Topič

## CRC Press
Taylor & Francis Group
Boca Raton London New York

CRC Press is an imprint of the
Taylor & Francis Group, an **informa** business

CRC Press
Taylor & Francis Group
6000 Broken Sound Parkway NW, Suite 300
Boca Raton, FL 33487-2742

First issued in paperback 2019

© 2013 by Taylor & Francis Group, LLC
CRC Press is an imprint of Taylor & Francis Group, an Informa business

No claim to original U.S. Government works

ISBN-13: 978-1-4398-1849-7 (hbk)
ISBN-13: 978-0-367-38020-5 (pbk)

---

**Library of Congress Cataloging-in-Publication Data**

---

Krc, Janez.
    Optical modeling and simulation of thin-film photovoltaic devices / authors, Janez Krc, Marko Topic.
        pages cm
    Includes bibliographical references and index.
    ISBN 978-1-4398-1849-7 (hardback)
    1. Solar cells--Mathematical models. 2. Thin film devices--Mathematical models. 3. Physical optics--Mathematics. I. Title.

TK2960.K737 2013
621.3815'42--dc23                                                                2013003040

---

**Visit the Taylor & Francis Web site at**
**http://www.taylorandfrancis.com**

**and the CRC Press Web site at**
**http://www.crcpress.com**

*Najinima staršema*

*(to our parents)*

# Brief Contents

# Preface

Photovoltaics is a young scientific discipline, and a fast-growing industry, which is already proving that it will significantly contribute to a sustainable supply of electricity. The photovoltaic (PV) technology mix of wafer-based silicon and thin-film technologies at the forefront is a solid foundation for future growth of the solar photovoltaic sector. It is believed that no single photovoltaic technology can satisfy the vast array of applications and consumer needs in different environments and installations, ranging from a few miliwatt stand-alone power supplies to multi-megawatt utility-scale power plants.

The production of thin-film PV devices has been expanding rapidly, particularly since 2005, when the production of thin-film solar PV modules exceeded 100 MW per annum. Tremendous growth in the following years was noted with 100 MW thin-film factories that became operational in 2007, followed by the first 1 GW factory in 2010. Low material consumption, large-area low-temperature deposition techniques and the possibility of using inexpensive flexible substrates have been the driving forces in research, development and implementation of thin-film photovoltaic devices in the market. In order to make thin-film solar cells and PV modules more competitive to conventional crystalline PV technologies, the conversion efficiencies of thin-film devices need to be further improved. A pivotal issue for thin-film PV devices in boosting their conversion efficiencies is efficient light management inside the structures. In this respect optical modeling and simulation have proven to be essential tools. They enable detailed analysis and accurate optimization of device structures. Moreover, they assist in testing and applying novel optical concepts in the devices. Modeling and simulation also possess a predictive power to study the potential improvements and to detect conversion efficiency limits of solar cells and PV modules.

This book is devoted to optical modeling and simulation of thin-film solar cells and PV modules. It was designed as a monograph and a comprehensive guide for performing optical modeling and simulations, giving insights into examples of existing optical models, demonstrating the applicability of optical modeling and pointing out concrete directions and solutions for improving the devices under scope. We have incorporated practical examples for using simulations with the developed models so that readers can better understand and develop their own models as well as innovative concepts in light management in thin-film photovoltaic devices. We discuss different approaches of one-, two- and three-dimensional optical modeling. Among the approaches we present in detail are one-dimensional (1-D) semi-coherent modeling and two-dimensional

(2-D) modeling based on the finite element method. The contents are aimed at a readership of post-graduate students and senior researchers who have an interest in optical simulation and light management of thin-film solar cells, as well as researchers in the industry. The technical level assumes that the reader has a knowledge of electromagnetic wave theory and basic principles of solar cells. Concepts and approaches presented here are in line with opportunities and challenges in thin-film PVs, although not limited solely to them. We sincerely hope that the concepts and approaches will be applied also in other fields of optoelectronics and photonics.

In the race for higher conversion efficiency of PV devices at the same or lower cost, optical design and light management prove to be an integral part of the optimization loop for any PV technology, either in laboratories or in production lines of solar cells and modules. Whether for wafer-based crystalline silicon PVs or third-generation emerging technologies, small improvements on a gigawatt scale of production mean millions of dollars or euros in savings.

The authors truly hope that this book will provide useful information and hints on optical modeling and simulations, encourage students and scientists to pay attention to and do modeling and simulations, and, last but not least, to highlight the significance of the field of research, development, and production.

# Acknowledgments

This monograph would not have been possible without our former outstanding PhD students and wonderful international cooperation, both formal (in many projects) and informal.

Sincere thanks go to Andrej Čampa and Benjamin Lipovšek from our lab who contributed many ideas and helped us to formulate various concepts and models, and carried out countless simulations above all.

We are deeply grateful to Miro Zeman and his team at TU Delft for continuous support in the fabrication of thin-film silicon devices and numerous fruitful discussions. We sincerely thank Christophe Ballif and his team at EPFL, Helmut Stiebig from Malibu GmbH & Co. and Bielefeld University, Milan Vaneček and Jiri Springer from the Czech Academy of Sciences, Ruud E. I. Schropp from Utrecht University and ECN, Marika Edoff from Uppsala University, João Conde from Instituto Superior Técnico, Wim Soppe from ECN, Makoto Konagai from Tokyo Institute of Technology, Michio Kondo from AIST Japan, Olindo Isabella from TU Delft, Martina Schmid from Helmholtz Zentrum Berlin, Jonas Malmström from Uppsala University, Oliver Kluth from Oerlikon Solar, Takuyi Oyama from Asahi Glass Company, Volker Hagemann from Schott, and others who supported us in the field of modeling and simulation.

We thank Luna Han, senior publishing editor at Taylor & Francis Group for the encouragement, patience, and careful checking of the materials for the book.

Last but not least, we gratefully thank our families, who followed our work with understanding and support.

# About the Authors

**Janez Krč, PhD,** is an associate professor in the Faculty of Electrical Engineering at the University of Ljubljana (Slovenia), where he is a member of the Laboratory of Photovoltaics and Optoelectronics. He received his PhD in electrical engineering from the University of Ljubljana in 2002. His research and teaching activities cover photovoltaics, optoelectronics, and analog electronic circuits. One of his main research interests is optical modeling and light management in thin-film solar cells.

**Marko Topič, PhD,** is a professor in the Faculty of Electrical Engineering at the University of Ljubljana, where he is the head of the Laboratory of Photovoltaics and Optoelectronics, and affiliate professor in the College of Natural Sciences at Colorado State University. He received his PhD in electrical engineering from the University of Ljubljana in 1996. His research interests include photovoltaics, thin-film semiconductor materials, electron devices, optoelectronics, electronic circuits, and reliability engineering.

He was a visiting researcher at Forshungszentrum Juelich, Germany, and a visiting professor at the University of Applied Sciences, Cologne, Germany, and at Colorado State University, Fort Collins, Colorado, United States.

Dr. Topič is the chair of the Slovenian Photovoltaic Technology Platform and a steering committee member of the European Photovoltaic Technology Platform. He is also a member of the Advisory Group on Energy in the Seventh Framework Programme of the European Commission.

He was the recipient of a research fellowship from the Alexander von Humboldt Foundation in 2000 and of the *Zoisova nagrada* (the highest award in the Republic of Slovenia for scientific and research achievements) in 2008.

# Principles

# Optical Properties and Modeling Approaches

**1**

CONTENTS

## 1.1   INTRODUCTION

The use of thin-film photovoltaic (PV) modules has been increasing continuously, and their share in the rapidly growing global market increases anywhere between 10% and 20% from year to year. In 2010, the annual production of thin-film PV modules reached 3.2 gigawatts (GW) of nominal power. Compared to the dominating wafer-based crystalline silicon PV modules, thin-film PV accounted for 15% of global PV module production in 2010 (Jäger-Waldau 2011).

### 1.1.1   ADVANTAGES AND CHALLENGES AHEAD

The general advantages of thin-film technologies over crystalline silicon are low-temperature deposition processes on large areas, lower material consumption, compatibility with some already established large-area technologies (such as glass or display industry), and the possibility of using low-cost flexible substrates (Street 2000). Thin-film PV technologies are focused on low-cost, high-conversion efficiency and environmentally stable products. Most thin-film technologies follow the concept of sequential large-area deposition of thin-film layers across the whole substrate combined with schemes of monolithically interconnecting solar cells in series (Luque and Hegedus 2011). This approach merges the cell and module production steps in the PV value chain. With the deposition of thin-film layers and devices on flexible substrates in roll-to-roll processes, roof-shingle-like connection of solar cells has become an alternative approach invented for different thin-film technologies and utilized by several companies (Uni-Solar, Flexcell Nanosolar, Solarion, Hyet Solar and others). In both cases, the conversion efficiencies of PV modules are lower than in that of small-area solar cells, but the gap between them is narrowing. Current world record conversion efficiencies of small-area solar cells for different thin-film PV technologies are given in Table 1.1.

To compete with conventional wafer-based silicon or III-V technologies in this respect, advanced improvement in conversion efficiencies of thin-film solar cells is still required. For example, monocrystalline silicon solar cells have reached an efficiency of 25%, whereas thin-film silicon solar cells are only above 12% stabilized efficiency. The efficiencies have to be increased to 20% to stay competitive with bulk crystalline technologies (Guha and Yang 2010).

---

**TABLE 1.1    Current Record Efficiencies of Different Thin-Film Solar Cell Technologies**

| Type of Thin-Film Solar Cell | Efficiency |
|---|---|
| CIGS | 19.6% |
| CdTe | 16.7% |
| a-Si:H | 10.1% |
| μc-Si:H | 10.1% |
| Tandem a-Si:H/μc-Si:H | 12.3% |
| Dye-sensitized | 11.0% |
| Organic (polymer) | 10.0% |

*Note:*   The values correspond to the record conversion efficiencies of laboratory small-area solar cells under standard test conditions (Green et al. 2012).

---

Another trend in thin-film technologies is to use thinner and thinner absorber layers. Absorber layers (or regions) are the layers (regions) where the absorbed light can generate pairs of free electrons and holes, which then can be extracted to the contacts and can contribute to the electrical current (photocurrent of the device). Thus, we want to absorb as much light as possible within the absorber layers of the cell and reduce absorption in other supporting layers in the structure. The trend of reducing absorber layer thickness is aimed at further minimizing material consumption (e.g., indium in CIGS absorber layer) and/or shortening the deposition times of layers (e.g., microcrystalline silicon absorber layer). To achieve the characteristics of high conversion efficiency on the one hand, and reduction of layer thickness on the other hand, research activities need to focus on optimization of existing solar cell structures and development of novel designs of thin-film solar cells and PV modules.

Improving both the optical and electrical properties of devices is important (Zeman et al. 2012). From the electrical point of view, low sheet resistance (high electrical conductivity) of the front contacts, an efficient ohmic junction between the oxide-based transparent contacts and doped semiconductor layers, decreased defect states in the region of charge carrier generation in absorber layers, and introduction of high-quality thin buffer layers at the interfaces are some ways to improve both open-circuit voltage and the fill factor of semiconductor thin-film solar cells (Shah 2010). From the optical point of view, we want to achieve sufficiently high absorption in the absorber layers, which are two to three orders of magnitude thinner (in the range of a few nanometers to a few micrometers) than in conventional solar cells (100–300 μm). Special optical schemes that enhance light trapping in the structures of thin-film PV devices have to be considered. This is especially important for the cells with indirect bandgap semiconductor absorbers (such as silicon) where the absorption coefficient of the material becomes low for the photons with energies close (but still above) the energy bandgap. Most commonly used light-trapping techniques in thin-film solar cells are established by using the effect of light scattering at textured interfaces in combination with high reflection at the back contact (reflector) of the cell (Schropp and Zeman 1998; Shah 2010). The area of designing new light-trapping techniques and optimizing existing ones is called *light* or *photon management* of thin-film solar cells.

Optical modeling and simulations have proven to be indispensable tools for designing and optimizing optical properties of materials and structures of thin-film PV devices. In this book we use the term *optical modeling* to refer to the development of models describing optical properties of thin-film layers, devices, and structures, whereas we assign the term *simulations* to employment of the simulators, which are based on these models, for the optical analysis of PV devices. As for *PV devices*, we consider solar cells and PV modules in particular.

The aim of optical modeling and simulations of solar cells is to help us to understand, design, and finally improve the optical behavior of the analyzed PV structures, resulting in higher conversion efficiency of PV devices. Moreover, modeling and simulations should also have a strong predictive power indicating the directions for future improvements of the devices and also detecting their realistic limitations.

Optical modeling and simulations are important for thin-film semiconductor, dye-sensitized, and organic solar cells. In the latter case, thin absorber layers and thus efficient light trapping are important because of short diffusion length of charge carriers in organic materials. In the case of dye-sensitized solar cells, one has to enhance absorption of long-wavelength light where the absorption coefficient of dye molecules becomes low. In this book we focus on thin-film semiconductor PV devices, although this does not mean that the models and approaches are not applicable to other types of thin-film solar cells.

## 1.1.2  ORGANIZATION OF THIS BOOK

The book is divided into two main parts: principles and applications. The first part deals with modeling and consists of four chapters, while the second addresses simulations and consists of two chapters.

In the first chapter, following this introductory section, we address some general aspects of optical modeling and simulations of thin-film PV devices. First, we briefly present basic properties of illumination sources that present excitation of our optical systems and must be considered as one of the input parameters in optical simulations. Then we highlight main optical features of thin-film devices that must be taken into account when developing the optical models. Afterward we briefly present principles and give an overview of different approaches of optical modeling. As a starting example of modeling of thin-film solar cells, we describe an approach to model multilayer systems of thin-film solar cells with only flat interfaces. In the last part of the chapter, we address main general steps in the procedure of optical modeling and simulations.

In Chapter 2 we present the development of a one-dimensional (1-D) semi-coherent optical model and its implementation in the Sun*Shine* simulator that can be used efficiently for simulations of different thin-film PV (and some other optoelectronic) devices with flat and nano-textured interfaces. Chapter 3 is devoted to description and determination of input parameters for optical simulations, which present an essential part in the simulation process. In Chapter 4 we highlight aspects of two- and three-dimensional (2-D and 3-D) modeling and as an example we present the development of a 2-D optical model and the simulator FEMOS, which are based on the finite element method of solving Maxwell's equations.

The second part of the book is devoted to applications of models in simulations. In Chapter 5 we demonstrate the applicability of the developed 1-D Sun*Shine* simulator on different cases, such as investigation of the role of light scattering at internal interfaces, antireflection effect at front interfaces, the distribution of light absorption in multi-junction solar cells, and other examples. In Chapter 6 we employ the 2-D FEMOS simulator to optimize antireflecting and light-scattering effects in thin-film silicon solar cells with periodically textured interfaces.

This book aims to give readers a sufficient basic understanding on how we can understand, approximate, and model main optical effects in thin-film structures of PV devices. With chosen examples of modeling and selected simulation results, we hope to establish sufficient knowledge, useful directions, and motivation for further research in the field of optical modeling and simulation of thin-film solar cells.

## 1.2    PROPERTIES OF LIGHT

### 1.2.1    REPRESENTATION OF LIGHT

Different models describing properties of light have been established by great scientists in the past, including Newton, Descartes, Huygens, Young, Maxwell, and others (Saleh and Teich 2007). For the purpose of this book we highlight two of the models that are nowadays commonly used in general optics. These are the *photon (particle) model* and the *wave model* of light (Smith, King, and Wilkins 2007). In case of the **photon model,** the propagation of light is represented by the flux of particles (energy quanta) called photons (Figure 1.1a). The energy of a single photon is determined by Equation (1.1)

$$E_{ph} = h \cdot \upsilon = h \cdot \frac{c_{medium}}{\lambda_{medium}} \tag{1.1}$$

where symbol $h$ is the Planck constant ($h = 6.626 \cdot 10^{-34}$ Js), $\upsilon$ is the frequency of light, $c_{medium}$ is the speed of light in the medium (in free space $c = c_0 = 299,792,458 \approx 3 \cdot 10^8$ m/s), and $\lambda_{medium}$ is the light wavelength in the medium which can be determined as $\lambda_{medium} = \lambda/n$ ($\lambda$ = light wavelength in free space, $n$ = refractive

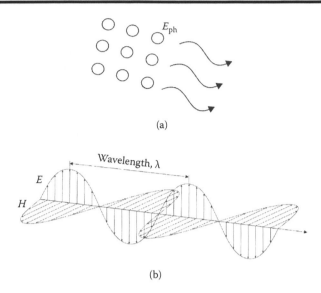

FIGURE 1.1    (a) Schematic representation of a photon flux (quantum model of light) and (b) of an electromagnetic wave (wave model of light) traveling through the space.

index of the material). The power density, $I$, of the photons of a single wavelength can be further determined as

$$I = \phi \cdot E_{ph} \tag{1.2}$$

where $\phi$ is the photon flux (number of photons per area per time). Usually photons of different energies, corresponding to different wavelengths [as shown in Equation (1.1)], are present in the illumination, forming the *illumination spectrum*. The total power density of the spectrum can be obtained by the summing (or integration) of $I$ over the entire wavelength region of the spectrum.

In the **wave model**, light is represented by electromagnetic waves. As shown later, Maxwell's equations present the basis for resolving the problems of light propagation and interaction with the material in this case. Vectors of the electric, $E(t)$, and the magnetic field strength, $H(t)$ ($t$ indicates time dependency of the quantities) and the direction of light propagation define the light at a certain spatial point in the wave model. In isotropic media vectors, $E$ and $H$ of an electromagnetic wave are perpendicular to each other and perpendicular to the direction of propagation of the electromagnetic wave, as shown in Figure 1.1b.

By using the Fourier transform we can convert our electromagnetic problem from the time domain to the frequency (wavelength) domain, which appears more convenient for the analysis (Saleh and Teich 2007). The real vectors $E$ and $H$ become complex vectors that are wavelength dependent. In this case, the power density of light, which we assign to a vector at this stage to include information about the direction of light propagation, $I$, of a certain wavelength at a certain spatial point, is defined by the Poynting vector [Equation (1.3)]

$$I = \frac{1}{2}\text{Re}[E \times H^*]$$    (1.3)

where $E$ and $H^*$ are the electric and conjugated magnetic field in the complex domain. In the equation, $I$ is a vector, because it includes the direction of light propagation determined by the vector product $E$ and $H$. In case we assume only forward and backward propagation of light, we often consider $I$ not as a vector but as a positive or negative scalar value. Further on, for simplicity we address the scalar power density of light $I$ as the *intensity of light*, although we have to keep in mind its original meaning. Also in the wave model of light, an integration of $I$ over the wavelengths gives the power density (intensity) of the entire spectrum. In sun illumination, this value is often assigned to 1000 W/m² (or 100 mW/cm²).

In advanced optical modeling approaches we often use the wave model of light. This model includes interference (coherence) effects of light, which are pronounced as interference fringes in measured spectral characteristics of realistic PV devices, as a consequence of forward and backward propagating waves in thin layers. Still, the photon model of light is not obsolete; it can be used successfully in combination with the wave model to describe the scattered light propagation in thin-film structures as a photon flux or intensity of the light beam (ray). For example, the photon model in combination with the Monte Carlo method (Keller, Heinrich, and Niederreiter 2007) can be used to determine the optical situation related to scattered light in some of the models (Springer, Poruba, and Vaneček 2004). Generally, all models based on rigorous solving of Maxwell's equations consider the wave and not the photon representation of light.

## 1.2.2    ILLUMINATION SOURCES

We can divide illumination sources into two categories: natural and artificial. The most obvious natural source is the sun, presenting the most common illumination in the case of PV devices. Artificial sources include different lamps (from blackbody radiation tungsten lamps to gas-discharge lamps) and semiconductor-based devices, such as light-emitting diodes (LEDs) or laser diodes (LDs) (Saleh and Teich 2007). Artificial sources are used in laboratories to characterize, optimize, and predict the behavior of thin-film PV devices under outdoor solar illumination.

One important characteristic of an illumination source is its spectrum. It presents the spectral distribution of irradiance, that is, the power density of incident light distributed across different wavelengths (photon energies). The spectrum is usually presented with units W/(m²μm), mW/(cm²nm), or similar. Examples of the measured spectrum of the sun on a clear day and on a cloudy day are shown in Figure 1.2.

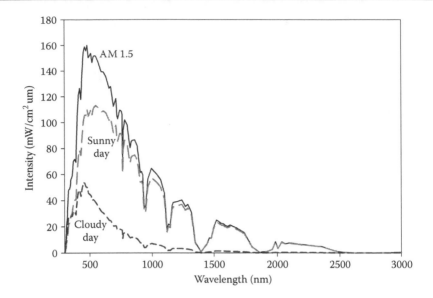

**FIGURE 1.2**   Spectra related to solar illumination, as measured on a sunny day and on a cloudy day (in June, Ljubljana, Slovenia), and the standard AM 1.5 solar spectrum (IEC 60904-3:2008).

For characterization of PV devices, we use the standard spectrum AM 1.5 with the total illumination intensity of 1000 W/m² (IEC 60904-3:2008; Luque and Hegedus 2011). The total illumination intensity can be calculated as an integral of the spectrum along the entire wavelength interval. The AM 1.5 spectrum is presented in Figure 1.2 in addition to the measured outdoor spectra. The solar spectrum on a clear sunny day approaches the AM 1.5 spectrum. If meteorological and geographical conditions are fulfilled, these two spectra match (Myers et al. 2000). Artificial sources are used to simulate the AM 1.5 spectrum indoors. In particular, so-called solar simulators, which consist of different types of lamps (e.g., xenon or halogen), are used to approach the AM 1.5 spectrum. Different classes of steady-state (constant light) and flasher simulators are available (Photon International 2002). With a solar simulator the main output parameters of a solar cell or PV module, such as short-circuit current, open-circuit voltage, fill factor, and conversion efficiency (Green 1981), are extracted from the current/voltage (*I-V*) characteristic determined under standard test conditions (STCs), which require matched illumination to the AM 1.5 solar spectrum and temperature of the measured device of 25°C (IEC 60904-3:2008).

Artificial illumination sources for characterization of PV devices can be divided further into broadband and narrowband sources. Broadband sources include solar simulators and most of the lamps that emit "white" light.

On the other hand, LEDs are narrowband sources in which the central wavelength is determined by the energy bandgap of the semiconductor used.

In combination with photoluminescent phosphor materials (such as photon energy downconverters), LEDs can also be made as broadband sources (e.g., white LED). Extremely narrowband sources, which can be assigned to monochromatic sources, are laser-based devices. A specific wavelength can be amplified and emitted from the laser through the use of resonators inside laser structures. Narrowband illumination sources are used to determine spectral responses of a PV device. To sweep the wavelength characteristic with a narrow band source, monochromator compartments in combination with a broadband source (wolfram, xenon, or halogen lamp) are used. In addition, gas lasers (HeNe) or LDs are employed if high-power density of monochromatic illumination is needed. LEDs, which in general have a relatively narrowband spectrum, are used as a bias light source in spectral response measurements. Examples of spectra of selected artificial sources are shown in Figure 1.3.

In optical simulations we describe the applied incident illumination (either natural or artificial) by a finite number of discrete wavelength intervals (e.g., 10-nm steps) of the spectrum. The intervals are represented by their central wavelength $\lambda_i$ and the corresponding power density in mW/cm² of the interval.

One important property of illumination that has to be considered in optical simulations of thin-film solar cells is *coherency* of light (Smith, King, and Wilkins 2007). Two or more coherent waves of the same wavelength propagating in opposite directions, or crossing each other, will interact. This means that at a certain point they can cancel or amplify each other, depending on their phase difference. As already mentioned, we have to use the wave model of light

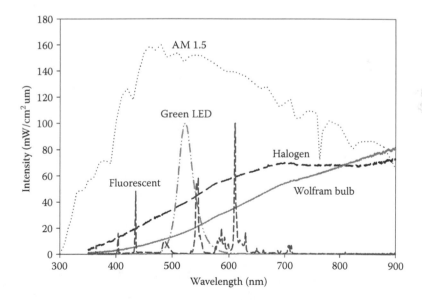

**FIGURE 1.3**  Illumination spectra of different artificial light sources and the AM 1.5 spectrum as a reference.

to include these coherency effects. In the model, a superposition of electric or magnetic field strengths is considered and not the superposition of light intensities, which can be used in case of incoherent propagation of light. We can observe the effects of interactions of coherent light waves as interference fringes in measured or simulated spectral responses of thin-film solar cells. The level of coherency of a source is defined by the coherency length. Solar illumination has a coherency length of around 1 μm (Born and Wolf 1999), and artificial sources used for characterization of solar cells have even longer coherency lengths. This means that with inside layers that have an effective optical thickness (geometrical thickness divided by the refractive index of layer) smaller than ~ 1 μm, coherent propagation of light has to be considered.

Another aspect that should be taken into account for solar illumination is the presence of *direct (specular)* and *diffused light* components. The direct component presents the light that comes directly from the sun, whereas the diffused component comprises the reflections, scattering, and diffraction of sunlight from different objects (clouds, terrestrial objects) Therefore, angular distribution of incident light has to be taken into account in simulations. Incident illumination into a 1-D optical model is discussed in the next chapter.

## 1.3   OPTICAL PROPERTIES OF THIN-FILM PHOTOVOLTAIC DEVICES

Photovoltaic devices are structures that are usually comprised of many layers. Table 1.2 gives the approximate number of layers that are typically present in different types of solar cells.

**TABLE 1.2   Typical Number of Layers in the Structures of Different Solar Cell Technologies**

| Type of Solar Cell | Typical Number of Layers |
|---|---|
| **Thin-Film** | |
| CIGS | 6–8 |
| CdTe | 5–7 |
| Thin-film silicon | 6–12 |
| Dye-sensitized | 6–8 |
| Organic | 5–7 |
| **Wafer-Based** | |
| Silicon | 3–10 |
| III-V concentrator double-junction | 12–15 |
| III-V concentrator triple-junction | 18–21 |

The number of layers included in an optical simulation can differ, depending on whether it is performed on the solar cell or at the PV module level. Here we should mention two basic configurations of thin-film solar cells: (a) superstrate (b) substrate configuration.

In the superstrate configuration, the substrate (carrier of the cell) is present at the front side where the light enters the structure. Thus, it has to be transparent, so glass plates are typically used for substrates (superstrates) in this configuration. No additional encapsulation front glass is needed for PV module fabrication in this case; only sealing with EVA (ethylene-vinyl acetate) or other adhesive foil and back sheet are added at the back (see Figure 1.4). Because no layers are added at the front side of the solar cell structure and since the light does not penetrate through the back contact to the two layers added at the back side, optical simulations at the cell and PV module levels are equivalent. In the second configuration the substrate is located at the back side of the cell. For module fabrication, encapsulation glass and EVA foil are added on the front side, affecting the optical system. Therefore, the simulation structure on the cell is different from that on the module level.

Detailed structures of thin-film solar cells are presented later in Sections 3.2.2, 3.5.2, and 5.3.1–5.3.4—at this point we consider only the solar cell and PV module structure as a general multilayer stack with 5 to 12 layers as specified in Table 1.2. The layers are usually semi-transparent (partially absorbing), except

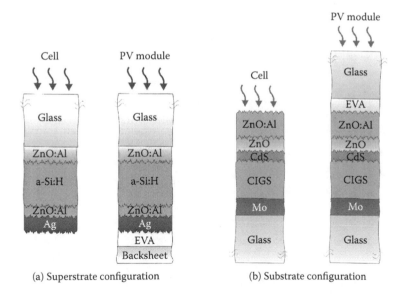

(a) Superstrate configuration                    (b) Substrate configuration

**FIGURE 1.4**   Differences in optical systems on solar cell and PV module level for (a) the superstrate (example of an a-Si:H-based PV device) and (b) the substrate configuration (example of a CIGS-based PV device).

that the back contact in most cases is opaque (metal). The light gets absorbed in the layers while propagating throughout. At the same time, it interacts with the interfaces of the multilayer structure. In an optical description of the complete device, optical properties of the materials employed as layers and optical properties of the interfaces have to be defined.

Optical properties of the materials are most often described by their complex refractive indexes, $N(\lambda) = n(\lambda) - jk(\lambda)$,* with $n$ as the real part (commonly called the refractive index or index of refraction) and $k$ as the imaginary part (called the extinction coefficient). Another possible description is by using complex dielectric (permittivity) functions $\varepsilon(\lambda) = \varepsilon_1(\lambda) - j\varepsilon_2(\lambda)$. The symbol $\lambda$ denotes the wavelength dependency of $N$ and $\varepsilon$, which typically cannot be neglected since the spectrum of interest in PV is relatively broad. The relation between $N(\lambda)$ and $\varepsilon(\lambda)$ follows the square root relationship [Equation (1.4)]:

$$\varepsilon_1(\lambda) = n^2(\lambda) - k^2(\lambda) \qquad \varepsilon_2(\lambda) = 2 \cdot n(\lambda) \cdot k(\lambda) \qquad (1.4)$$

$\varepsilon_1$ and $\varepsilon_2$ are relative dielectric functions. Absolute dielectric are obtained by multiplying them by $\varepsilon_0 = 88542 \times 10^{-12}$ F/m. Besides dielectric functions, which relate to the electric field, permeability properties of the material, which relate to the magnetic field, present important electromagnetic characteristics. However, in the range of optical frequencies, relative permeability of the materials used in thin-film PV devices can be assumed to be unity, its value is assumed to be one.

Because the layers in PV devices are mostly isotropic and more or less homogenous, $N$ and $\varepsilon$ are not vectors or tensors. However, materials used are partially absorbing; therefore, $N$ and $\varepsilon$ are complex functions in general (see examples in Chapter 2). In further studies, we will use mostly $N$ for the description of optical properties of layers. The real part $n(\lambda)$ determines the light wavelength in the material, as already specified in Equation (1.1). The imaginary part $k(\lambda)$ (the wavelength dependency of $n$ and $k$ will no longer be written in further descriptions) determines the absorption of light in a layer and is linked to the absorption coefficient $\alpha$, which is more generally used for definition of absorption in the material, by Equation (1.5).

$$\alpha = \frac{4\pi k}{\lambda} \qquad (1.5)$$

The decrease of light intensity when propagating throughout an absorbing medium can be given by the Beer–Lambert formula (Ingle and Crouch 1988):

$$I(x) = I_0 \cdot e^{-\alpha \cdot x} \qquad (1.6)$$

---

* In some literature, the imaginary part of complex refractive index or dielectric function has a positive sign. Equations where the quantities are used have to be changed accordingly.

where $x$ is the axis in the direction of light propagation and $I_0$ is the light intensity at the starting point where $x = 0$.

When the light hits an interface between two layers, sudden change in the optical properties of the material affects its propagation. In general, reflection of light in a backward direction and transmission into the next layer in a forward direction occur. In textured interfaces, we have to consider the effect of light scattering, which plays an important role in light management of thin-film solar cells. Light scattering introduces multiple new directions of propagation of reflected and transmitted light. Here, let us define the reflectance and transmittance coefficients* of light for a simple case of a flat interface and perpendicular illumination of light. On the level of light intensities the reflectance, $R_{12}$, and transmittance, $T_{12}$, are defined as in Equations (1.7) and (1.8). The numbers in the subscripts correspond to denotation of the incident medium/layer (1) and the medium/layer in transmission (2)

$$R_{12} = \left| \frac{N_1 - N_2}{N_1 + N_2} \right|^2 \tag{1.7}$$

$$T_{12} = 1 - R_{12} = 1 - \left| \frac{N_1 - N_2}{N_1 + N_2} \right|^2 \tag{1.8}$$

In the equations, $N_1$ and $N_2$ are wavelength-dependent complex refractive indexes of the incident layer and the layer in transmission, respectively. We can see that the bigger the difference is between the complex refractive indexes, the greater the reflectance and the lesser is the transmittance. Thus, in an interface between materials with large contrast in complex refractive indexes, a significant amount of light is reflected. The equation for $T$ follows the energy conservation law at an interface ($R + T = 1$).

So far we have become acquainted with basic descriptions and definitions of some optical properties of layers and interfaces that must be considered in optical modeling of thin-film PV devices. We also have defined the attenuation of intensity of a light ray (photon model behind) propagating throughout a (semi) absorbing layer and reflectance and transmittance for perpendicular incidence of light at a flat interface. Also in a solar cell structure we have to consider (a) propagation of light and (b) the situation at interfaces. However, the situation is not so trivial since light is propagating in forward and backward directions as a result of multiple reflections and transmissions at the interfaces. Further on,

---

* We relate the terms *reflection* and *transmission* to general phenomena that occur when we shine the light at an interface, whereas we assign the terms *reflectance* and *transmittance* (coefficient) to specific quantities determining how much light is reflected and transmitted (on the level of electric field or light intensity). The value of these quantities can be between 0 and 1 or between 0% and 100%, depending on the representation.

as mentioned, in the case of thin-film solar cells, textured interfaces are usually introduced, making the analysis much more complex. We have to employ so-called advanced approaches of optical modeling.

To summarize, the main optical effects in thin-film PV devices that have to be considered in an optical model are

light propagation throughout the layers (including interactions of coherent waves),
reflection and transmission processes at interfaces, and
light scattering at textured interfaces or introduced particles.

All three effects are discussed and classified in more detail in the next chapter.

## 1.4   APPROACHES OF OPTICAL MODELING

Optical models are based on equations describing the physics of optics in the analyzed structures. In the past it was of prime importance that the equations or resulting system of equations could be solved analytically. This was possible for simple optical systems and often conditioned with the use of certain simplifications of the system. Nowadays, in the era of supercomputers, numerical approaches dominate. We can divide optical modeling into the following basic steps (see also Section 1.6):

Define the equations that describe optical situation (device optical properties).
Choose a numerical approach to solve the problem (system of equations).
Implement the model in a computer code (simulator).

The computer code is usually upgraded with a graphical user interface, to set the input parameters and settings of simulations and to display the obtained results easily and immediately. Such a complete computer program represents the main core of a simulator.

Depending on dimensions and optical properties of the layers and interfaces in an optical system and properties of the illumination, different optical models have been developed. One classification is related to the number of spatial dimensions (one, two, or three) the model considers; thus, we have 1-D, 2-D, and 3-D models. As discussed in Chapters 2 and 4, in thin-film PV devices 1-D models can be applied quite successfully in many cases.

The second model classification that we mention here is based on the relationship between dimensions of the structure with respect to light wavelength. Here, for dimensions of the structure we mean thicknesses of layers, vertical and lateral parameters of surface textures, size and distance between the particles (if introduced in the structure), and other spatial variations of refractive index in the structure. If the dimensions of the structure are much larger than the wavelength of light, optical rules defined for geometrical optics can be used in the modeling. We assign the models that are based on geometrical optics as *macro optical models*. In these models, light can be represented by their intensity or photon flux; no phase information of the electromagnetic waves is needed. The propagation of light is described by tracing of rays, considering absorption in the layers defined by absorption coefficient of the material and reflections/transmissions at the interfaces, which are determined by the complex refractive indexes of adjacent layers and the angle between the direction of the incident ray and the plane of the interface (or individual segment in the case of textured interfaces). Details of these are given in Chapter 2. Different ray tracing models have also been developed for PV applications (see, e.g., Cotter 2005; Schulte et al. 2011; ASAP numerical program; Synopsys).

In Figure 1.5 a schematic representation of ray tracing in textured glass is presented. Light rays are applied from the top side (incident medium air) and scattered at a textured glass surface, following geometrical optics. The figure was created by means of CROWN simulator (Lipovšek et al. 2011).

The next category of optical models covers models that describe systems in which the dimensions of structures are much smaller than the wavelengths of applied illumination. We call these optical models *nano models* or *micro models*. Here, electromagnetic propagation of light throughout the layers has to be considered, since the interaction between the waves and their phase shifts becomes important. For the surface textures where vertical parameters, such as root-mean-square roughness, are much smaller than the wavelength of light, different scalar theory

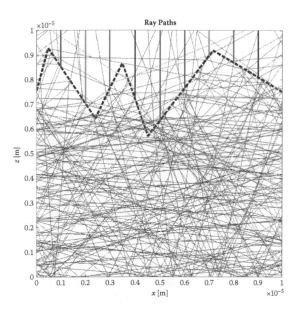

**FIGURE 1.5**    Schematic representation of ray tracing procedure (top medium air, bottom layer glass, light is scattered at textured interface, (represented by dashed line) resulting in propagation of rays in different directions).

approaches are employed to determine the effects of the texture on the reflectivity and transmission properties of the interface (Beckmann and Spizzichino 1987). Due to small roughness, the level of scattered light is low and specular light dominates. For example, *effective medium theory* (Choy 1999) and different *scalar scattering theories* (Carniglia 1979; Beckmann and Spizzichino 1987; Born and Wolf 1999) and *vector scattering theories* (Elson and Bennett 1979; Tsang, Kong, and Ding 2000) can be applied to such textures under certain conditions to determine the antireflection effect and the effect of light scattering at the textured interfaces, respectively. Some of these approaches are addressed in Chapter 3.

The last category of optical models in the classification based on spatial dimensions of the system with respect to the wavelength of light are models that can be used to describe the optical situation in structures where the dimensions are in the range of light wavelengths. For this complicated optical situation, often exact solutions to the problem cannot be found analytically. Combinations of different theories and models are applied to find a good approximation of the realistic situation. Therefore, we call the models describing optical situations in such structures *hybrid models*. Light propagation throughout the layers with the thicknesses in the range of light wavelengths is similar to what we see in nano- or micro-optical models; thus, the wave nature of light has to be considered. Scalar scattering approaches have been applied to analyze optical effects related to the textures; lateral and vertical dimensions in the range of light wavelengths are extended and empirically calibrated to approximate the optical situation in such structures (Stiebig et al. 2000; Zeman et al. 2000; Krč et al. 2002, 2003). Mie solutions of Maxwell's equations also can be effectively used in this case for describing light scattering on particles as well as on textured interfaces (Wriedt and Hergert 2012). In addition, vector scattering theories and different perturbation approaches (describing the changes in surface morphology as perturbations) have been employed (Elson and Bennett 1979; Elson, Rahn, and Bennett 1980), although the parameters of the texture often exceed the validity conditions.

On the other hand, the exact solution of such optical systems can be obtained by using so-called rigorous models, where first principles are followed by directly solving Maxwell's equations in the structure (Jin 2002). In their differential form, Maxwell's equations are given as

$$\nabla \cdot \boldsymbol{E} = \frac{\rho}{\varepsilon}$$

$$\nabla \cdot \boldsymbol{B} = 0$$

$$\nabla \times \boldsymbol{E} = -\frac{\partial \boldsymbol{B}}{\partial t} \tag{1.9}$$

$$\nabla \times \boldsymbol{B} = \mu \boldsymbol{J} + \mu \varepsilon \frac{\partial \boldsymbol{E}}{\partial t}$$

where ρ is space charge density, $\boldsymbol{J}$ is the current density related to electromagnetic oscillations, and symbols ε and μ are complex permittivity (or complex dielectric function) as introduced in Equation (1.4) and permeability of the medium. Wave equations are derived from Maxwell's equations (see Section 4.3.1) and are more suitable for solving electromagnetic wave problems.

In principle, this fundamental electromagnetic approach would be applicable to any optical system because it follows the fundamentals of electromagnetic wave propagation and the interaction between waves and the material. However, the available numerical methods require relatively dense discretization of the analyzed structure (sub-wavelength discretization), which can lead to an enormous number of points (nodes) in which these equations should be solved. The full extent of the problem arises if the equations are solved in the structures where dimensions in the range of millimeters have to be considered (i.e., three or more orders larger than the wavelength of light). The size and complexity of the resulting system can quickly become too large to be solved directly or iteratively in a timely manner, even with today's supercomputers. Therefore, rigorous approaches are more suitable for determining the optical situation in smaller domains, if possible, where the structures are up to a few cubic micrometers (e.g., closely around textured interfaces). Periodicity or symmetry properties of the structures have to be considered in such calculations, which can minimize the domain of simulation drastically. In Figure 1.6, an example of calculated light intensity in an a-SiC:H layer with periodically textured surface is shown. Here, the incident light was applied in the form of a Gaussian beam. Scattering in discrete orders can be nicely seen from simulations.

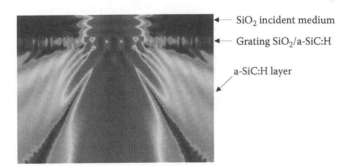

SiO$_2$ incident medium

Grating SiO$_2$/a-SiC:H

a-SiC:H layer

**FIGURE 1.6**  Intensity of light scattered at a periodically textured interface between SiO$_2$ (top) and the a-SiC:H layer (bottom) as calculated by a rigorous solving approach (using the finite element method) (Campa, Krč, and Topič 2009). The period and the height of the periodic texture are $P = 160$ and $h = 170$ nm, respectively. The wavelength of applied illumination (perpendicular incidence from top side) is $\lambda = 400$ nm, transverse electric (TE) polarization.

Different numerical methods of rigorously solving the equations have been applied in optical modeling of thin-film solar cells. Most common are

- finite difference time domain,
- finite element method,
- finite integration technique, and
- rigorous coupled-wave analysis.

The finite difference time domain (FDTD) is a widely used method to solve electromagnetic problems (Taflove and Hagness 2005; Elsherbeni and Demir 2009). A wide range of frequencies (wavelengths) can be included in one simulation run since the method works in a time domain. The electric field components in a volume of space are solved at a given instant in time; the magnetic field components in the same spatial volume are solved at the next instant in time; and the process is repeated over and over again until the desired transient or steady-state electromagnetic field behavior is fully evolved. The approach was also applied to investigate the optical situation in thin-film solar cells (Dewan et al. 2011; Lacombe et al. 2011; Pflaum and Rahimi 2011; Solntsev and Zeman 2011). One disadvantage of the approach that we experienced is that complex refractive indexes of layers, which are usually given as wavelength-dependent functions, have to be represented by dielectric functions and then parameterized with Lorenz oscillators (Oughstun and Cartwright 2003). At shorter wavelengths of the solar spectrum, where dielectric functions of semiconductor materials have noticeable wavelength dependency, more than three oscillators have to be superimposed to reassemble the wavelength behavior of dielectric functions. This often leads to convergence problems in finding solutions or at least in long computation times. Another difficulty one can experience with FDTD in the analysis of thin-film photovoltaic devices is related to metal back contacts. When realistic optical properties of metals such as silver or aluminum are used in simulations, convergence problems become of prime importance because of negative permittivity of the material. However, solutions have been found to overcome these problems, making the FDTD method fully applicable for simulations of thin-film solar cells (Dewan et al. 2011; Lacombe et al. 2011; Pflaum and Rahimi 2011; Solntsev and Zeman 2011).

The finite element method (FEM) is a numerical technique for finding approximate solutions of Maxwell's equations presented in their partial differential or integral form (Jin 2002). The solution approach is based either on eliminating the differential equation completely (steady state problems) or on rendering the partial equations into an approximating system of ordinary differential equations. These ordinary equations are then numerically integrated using standard techniques such as Euler's method, the Runge-Kutta method, and others. The FEM is a good choice for solving the electromagnetic situation over complicated structures. The method appears to be very effective also for determination of optical situation in thin-film PV devices (Čampa, Krč, and Topič 2009;

Isabella et al. 2012). Wavelength-dependent complex refractive indexes can be directly used as input parameters, which is convenient. Optical effects occurring at corrugated metal surfaces or metal nano-particles can be sufficiently solved with FEM. Therefore, this method was also chosen for development of the 2-D optical model presented in Chapter 4.

The finite integration technique (FIT) is a spatial discretization scheme to numerically solve electromagnetic field problems in time and frequency domain (Jin 2002). The basic idea is to apply Maxwell's equations in integral form to a set of staggered grids. This method stands out due to high flexibility in geometric modeling and boundary handling as well as incorporation of arbitrary material distributions and material properties such as anisotropy, non-linearity, and wavelength dependency. Furthermore, the use of a consistent dual orthogonal grid (e.g., Cartesian grid) in conjunction with an explicit time integration scheme (e.g., leapfrog scheme) leads to memory-efficient algorithms. Successful use of the method in simulations of thin-film silicon solar cells can be found in Haase and Stiebig (2006, 2007).

Rigorous coupled-wave analysis (RCWA) is a rigorous computational method for solving the electromagnetic modes in periodic dielectric configurations (Nevière and Popov 2002). To solve the electromagnetic mode in the periodic dielectric medium, Maxwell's equations (in partial differential form), as well as the boundary conditions, are expanded by Floquet functions and turned into infinitely large algebra equations. With the cutting off of higher order Floquet functions, depending on the accuracy and convergence speed we need, the infinitely large algebra equations become finite and thus solvable by computers. RCWA has been applied efficiently in simulations of thin-film silicon solar cells with periodically textured interfaces (Chen, Wang, and Li 2010; Zhao et al. 2010).

Rigorous methods of solving Maxwell's equations are usually used to determine optical situations directly from the geometry of textured interfaces in solar cell structures. Typically, lateral and vertical dimensions of the introduced textures in thin-film silicon solar cells are in the range of light wavelengths; thus, hybrid models and rigorous methods must be used. Among them, simulations based on rigorous methods are much more time consuming; therefore, well-calibrated hybrid models are still of interest. In addition to the hybrid model presented in the next chapter, two other hybrid models have been successfully employed in the analysis of thin-film silicon solar cells with flat and textured interfaces: the Genpro 3 model from Technical University of Delft, which is implemented in the simulator ASA (Zeman et al. 1997; Pieters, Krč, and Zeman 2006), and the cell optical model developed at the Czech Academy of Science in Prague (Springer, Poruba, and Vaneček 2004). Both models consider coherent propagation of direct light in thin layers and scattering at each textured interface. The Genpro 3 is based on transfer matrix formalism of describing wave and ray propagation in the structure. In the cell model the non-scattered coherent part of light is treated iteratively, whereas for tracing the scattered rays the Monte Carlo method is applied (Keller, Heinrich, and Niederreiter 2007). Other hybrid models for the analysis of thin-film solar cells have been reported as well (see, e.g., Lanz et al. 2011).

## 1.5 SIMPLIFIED APPROACHES TO OPTICAL MODELING OF THIN-FILM PHOTOVOLTAIC DEVICES

For a better understanding of the optical situation in thin-film PV devices, we first introduce and explain two simplified 1-D optical models of thin-film solar cells: the *classical* and the *extended classical* optical models. Then, we address a third model, the *wave optical* model, which is explained in more detail in the next chapter. In all three models, a thin-film solar cell is represented by a multilayer structure of semi-transparent layers, and the interfaces between the layers are assumed to be perfectly flat and plan parallel (see Figure 1.7).

As mentioned, the classical and the extended classical models present a simplification of the actual optical situation in the structure. However, despite the simplifications, useful results can be obtained, especially with the extended model. Let us focus first on the classical optical model.

### 1.5.1 CLASSICAL OPTICAL MODEL

In this model we consider the following optical properties and effects in the structure (Figure 1.7):

- Reflectance of light in the structure is employed only to the front surface, $R_{\text{front}}$, where light enters the structure, and to the back-layer front interface, $R_{\text{back}}$ (back layer is usually metal contact); reflectances at all other interfaces are ignored.

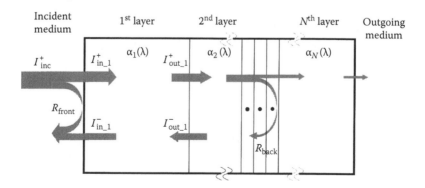

**FIGURE 1.7**   Schematic representation of optical situation in the classical optical model.

■ Absorption of light in each layer is determined according to the wavelength-dependent absorption coefficient $\alpha(\lambda)$.

■ Light is applied only perpendicularly to the front surface of the structure and is assumed to be fully incoherent; thus, it is analyzed in terms of incoherent rays, which are defined by their intensity and direction (forward, backward).

In this model the reflectances $R_{front}$ and $R_{back}$ are wavelength-independent parameters (constants), which can be set to a value from zero to unity. The two reflectances should represent an average of the actual wavelength-dependent reflectances of the corresponding interfaces, which are determined by the complex refractive indexes of layers [see Equation (1.7)]. Furthermore, $R_{front}$ can accommodate the reflectances of all internal interfaces, not just of the front surface. To keep the model simple, these two values are set to $R_{front} = 0.05$–$0.1$ and $R_{back} = 0.9$–$1.0$, without any extensive calculation. The reflected light intensities at these two interfaces, $I^-_{R\_front}$ and $I^-_{R\_back}$, are calculated as

$$I^-_{R\_front} = R_{front} \cdot I^+_{inc} \qquad I^-_{R\_back} = R_{back} \cdot I^+_{back} \qquad (1.10)$$

where $I^+_{inc}$ presents the intensity of incident illumination and $I^+_{back}$ is the intensity that reaches the back interface (for a specific discrete wavelength interval). The superscripts $+$ and $-$ stand for forward and backward propagation of light, respectively. The forward propagation is defined by the direction of applied incident illumination.

Light absorption in an individual layer is defined in the model by the wavelength-dependent absorptance parameter $A_{layer}$, which is calculated as

$$A_{layer} = \frac{I_{in} - I_{out}}{I^+_{inc}} \qquad (1.11)$$

where $I_{in}$ and $I_{out}$ are the common light intensity entering and leaving the layer, respectively. They account for both forward- and backward-going light intensity and are calculated as

$$I_{in} = I^+_{in} - I^-_{out} \qquad I_{out} = I^+_{out} - I^-_{in} \qquad (1.12)$$

The relationships between $I^+_{in}$ and $I^+_{out}$, and $I^-_{in}$ and $I^-_{out}$, are defined by the propagation of light throughout the layer as

$$I^+_{out} = I^+_{in} \cdot e^{-\alpha \cdot d} \qquad I^-_{out} = I^-_{in} \cdot e^{-\alpha \cdot d} \qquad (1.13)$$

where $\alpha$ is the absorption coefficient of the layer (for the calculated wavelength) and $d$ is the thickness of the layer. The given exponential decay in intensities of rays is schematically illustrated in Figure 1.8.

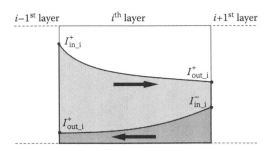

**FIGURE 1.8** Light intensities of forward (+) and backward (−) rays that enter (in) and leave (out) the *i*-th layer in the classical optical model. Exponential decay in the layer is determined by Equation (1.6).

The calculation procedure in the classical model is relatively simple. One sweep of light intensity in the forward and one in the backward direction are needed to determine $I^+_{in}$, $I^+_{out}$, $I^-_{in}$, and $I^-_{out}$ for all the layers. This enables calculation of absorptances in each layer and the total reflectance from the structure, which is defined as

$$R_{tot} = \frac{I^-_{out\_first\_layer} + R_{front} \cdot I^+_{inc}}{I^+_{inc}} \tag{1.14}$$

As we see, $R_{tot}$ consists of the component of light representing the light coming back from the cell and the reflected light at the front surface. Examples of results obtained with this model are shown and compared to other two modeling approaches at the end of this section.

### 1.5.2   EXTENDED CLASSICAL OPTICAL MODEL

In contrast to the classical model, in the extended classical model, wavelength-dependent reflectances and transmittances are considered at each interface of the structure. They are applied for both forward- and backward-going rays at an interface, resulting in multiple forward- and backward-going rays (Figure 1.9).

The reflectances and transmittances are calculated based on complex refractive indexes of the layers [Equations (1.7) and (1.8)], which here are input parameters, besides layer thicknesses and absorption coefficients. The light is still considered to be fully incoherent. The propagation of a light ray inside a layer is calculated in the same way as in the classical model, following the exponential decay determined by the absorption coefficient. However, in this case not only one forward- and one backward-going ray are present in the structure, as in the classical model. The situation is more complex because of multiple reflection and transmission processes at the interfaces. In this model, we apply a calculation algorithm that

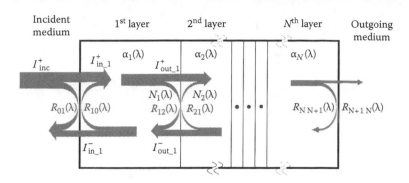

**FIGURE 1.9**   Schematic representation of optical situation in the extended classical optical model.

accounts for tracing of the rays in the structure. Different approaches can be used, such as transfer matrix formalism (Troparevsky et al. 2010) or the methods of individual ray tracing (Cotter 2005). One of them is presented in the next chapter in the frame of a semi-coherent optical model. The extended classical model would fully describe the actual situation in a thin-film solar cell structure with flat interfaces if the incident light had fully incoherent properties. However, as mentioned, solar illumination and most artificial illumination sources possess a certain degree of coherency; thus, the results obtained with the extended classical model are still approximations (as shown in Figure 1.11).

### 1.5.3   WAVE OPTICAL MODEL

To consider coherent nature of illumination, a model that takes propagation and interaction of light in terms of electromagnetic waves is needed. We call this model the wave optical model.

In contrast to the classical and extended classical models, the propagation of light and the reflection and the transmission process at each interface is defined on the level of electric (or magnetic) field strength (Figure 1.10). We consider superposition of the field vectors of forward- and backward-going waves of light at a specific spatial point in the structure. Reflectances and transmittances at an interface are determined for electric field rather than light intensities, to preserve the coherent character of light. The Poynting vector [Equation (1.3)] is used to determine intensities of light after the field has been calculated in the structure. The wave optical model presents the basis for calculation of coherent light propagation in the structure in the semi-coherent approach that we explain in detail in the next chapter; here we only addressed the main features of the model. This model describes the optical situation in thin-film structures with flat interfaces in a realistic way, as verified by optical measurements of fabricated structures.

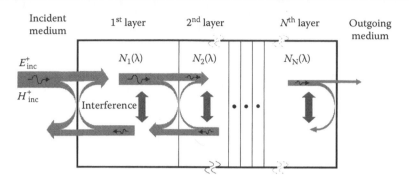

**FIGURE 1.10**   Schematic representation of optical situation in the wave optical model.

### 1.5.4   COMPARISON OF THE APPROACHES

In Figure 1.11a we show simulation results of a selected thin-film structure obtained by the classical, extended classical, and wave optical models. The simulations refer to an amorphous silicon (a-Si:H) solar cell structure that consists of six layers: glass, transparent conductive oxide (TCO) as front contact, $p$-doped, intrinsic, and $n$-doped hydrogenated amorphous (a-Si:H) layer, and metal (Ag) back contact. All interfaces are flat. At this point we will not go into details of the structure and its operational principle; we simply use it as an example to demonstrate the performance of the three previously mentioned models. In addition, thick layers such as glass (thickness in the range of millimeters) require special consideration in the wave model, which is explained in the next chapter. In Figure 1.11a we show the absorptance as a function of light wavelength calculated for the intrinsic layer with the thickness of 450 nm in the analyzed structure. When using different models, deviations can be observed. In the case of the absorptance curve obtained with the wave optical model, we can observe interference fringes (oscillations), which are a consequence of interactions of forward- and backward-going waves (coherent effects of amplifications and cancelations of the waves with respect to their phase). In the simulation obtained with the extended classical model, an averaged (flat) behavior can be observed, which is a consequence of incoherent treatment of light in this model. We can see similar flat curve behavior in the simulation that corresponds to the classical model. Here, the level of the absorptance is higher, since the reflectances at internal interfaces—especially at the front side of the structure—are ignored (underestimated in $R_{front}$). For comparison, the spectral dependent quantum efficiency, $QE$, of the realistic a-Si:H solar cell with flat interfaces* is plotted in

---

* In realistic thin-film solar cells, the interfaces are not perfectly flat; rather, they always have a certain roughness, ranging from 5 to 10 nm of vertical root-mean-square roughness.

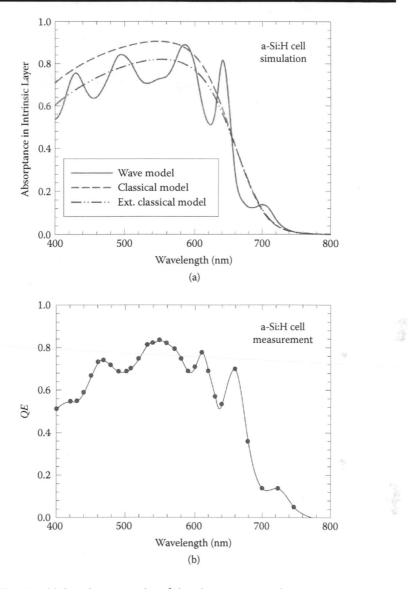

**FIGURE 1.11**    (a) Simulation results of the absorptance in the intrinsic (absorber) layer of an a-Si:H single-junction *p-i-n* solar cell as obtained with the three different models. (b) Measured external quantum efficiency of the cell.

Figure 1.11b. As explained later, the *QE* curve under certain conditions can be directly compared to the absorptance in the intrinsic layer of silicon thin-film solar cell. Interference fringes are observed in the measured *QE*, as in the case of simulation with the wave model. This confirms that the wave model is suitable for simulation of thin-film solar cells with flat interfaces, whereas the classical and extended classical models provide only a certain approximation of the optical situation and output performance of the thin-film structure.

Usually thin-film solar cells have (nano-)textured interfaces, where incident light is partially scattered. Advanced optical modeling approaches presented in Chapters 2 and 4 have to be applied to properly consider the effects of light scattering and propagation of scattered and non-scattered (direct) light in the structure. The models presented thus far (classical, extended classical, and wave) cannot be directly applied for simulations of such structures because they consider only propagation of direct light.

## 1.6  STEPS IN DEVELOPING AN OPTICAL MODEL AND SIMULATOR

Before developing more complex models like those presented in the following chapters, let us take a more general, systematic view of the process of modeling and simulation. The last example in the previous section showed that modeling and simulation go hand in hand with experimental work, not only to verify the final results of simulations but also to properly describe the realistic optical situation in the structure. This can be obtained by using suitable optical models and, very important, realistic input parameters in simulations. Well-determined realistic input parameters, such as complex refractive indexes and thicknesses of layers (and other parameters introduced later), are of prime importance for reliable simulations. Therefore, Chapter 3 is devoted to this topic. In this section we want to elaborate on the importance of all elements that need to be considered in the process of modeling and simulation (see Figure 1.12).

For development of reliable optical models, we need a well-established theoretical background in the field of electromagnetics and optics and knowledge about the actual optical properties and situation in specific device structures (first steps in the diagram in Figure 1.12). The second issue is also closely related to experimental work (see arrow from optical characterization to device properties). In particular, measuring and analyzing situations in simplified partial structures of the devices can give important indications of the optical effects that need to be included in the model. For example, indication about coherent behavior of light in thin-film structures, which needs to be taken into account, can be clearly identified already with a properly designed simple structure consisting of a single thin film on a transparent substrate. The wavelength-dependent measurements of reflectance or transmittance of the sample reveal interference fringes which can be, by proper physical understanding, related to the coherent nature of light. As explained in the introduction section of this chapter, an optical model is represented by equations, systems of equations, and algorithms, defining optical situation. One of the challenges in the development of an optical model is to make it as simple as possible (avoiding unnecessary effects) and closely related to the main principles of physics, to understand and

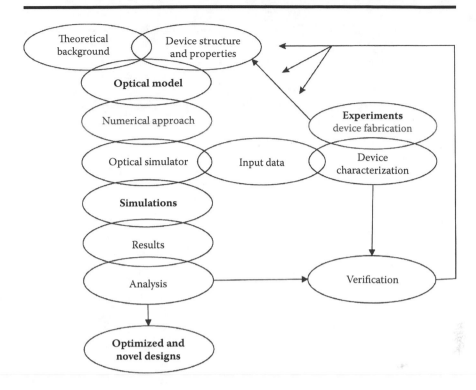

**FIGURE 1.12**    Main steps in optical modeling and simulations. Synergy between modeling and simulation and experiments is of prime importance for realistic simulations.

properly interpret the results obtained with simulations. There is still a need for developing efficient and accurate optical models that describe the optical situation in thin-film PV devices, especially in relation to the effects of light scattering at textured interfaces and particles.

Once the first version of the model is developed, we must choose a method for solving the equations and algorithms. In most cases, we pick a suitable numerical approach for whatever mathematical problem is defined by the model. The choice of the method is important from the point of view of convergence (toward the right solution), reliability and accuracy of the results, and computational time (a very important factor). We can find different numerical methods and recommendations on when to use certain methods (Jin 2002).

Once the numerical method is chosen, the model can be implemented in a computer code, which presents the core of a numerical simulator. Here, the most difficult decision is related to the selection of the hierarchical level of the computer language we want to use for implementation of our model. In general, lower level languages (such as C++) require more effort to describe our model equations and related calculations; however, the strong advantage of these languages is usually shorter computer run times. Higher level languages (such as MATLAB® or *Mathematica*®), on the other hand, enable simple implementation of the model but may suffer from longer run times. Since computational time is one of the main priorities, lower level languages are often used. However, higher level languages can now compete since they have been extensively upgraded as well as optimized for speed of calculation and memory consumption. For example, programming a model in MATLAB is a good choice.

Once the numerical simulator has been developed, we have to determine test input parameters to run the simulations and obtain the results. We describe determination of input parameters extensively in Chapter 3. To obtain realistic input parameters, experimental structures have to be fabricated and characterized. Results of optical characterization are compared with the simulation results. Analysis of deviations between the measured and simulated results directs us back to different aspects of model development (see multiple arrows in Figure 1.12). The theoretical bases of the model, numerical method, or input parameters have to be corrected or upgraded to approach the experimental results. Is this a never-ending story where we cannot end the iterative loop? Of course not. At this point, we have to point out that perfect fit to one experiment is much less important than good general agreement between simulated and measured curves corresponding to more experimental tests in which we vary the same parameter in experimental samples and simulations.

Once the model and simulator have been sufficiently verified, we can use them for optical optimization of existing solar cell structures and for design of novel structures. Nonetheless, one of the primary missions of optical modeling and simulations is their power of prediction, as already mentioned. This power can be used to establish reliable directions for improvements and new breakthroughs in the optical design of thin-film PV devices.

## REFERENCES

ASAP numerical program. http://www.breault.com/software/asap.php.
Beckmann, P., and A. Spizzichino. 1987. *The Scattering of Electromagnetic Waves from Rough Surfaces*. Artech Print on Demand.
Born, M., and E. Wolf. 1999. *Principles of Optics: Electromagnetic Theory of Propagation, Interference and Diffraction of Light*. 7th ed. Cambridge University Press.
Čampa, A., J. Krč, and M. Topič. 2009. "Analysis and Optimisation of Microcrystalline Silicon Solar Cells with Periodic Sinusoidal Textured Interfaces by Two-Dimensional Optical Simulations." *Journal of Applied Physics* 105 (8) (April 15). doi:10.1063/1.3115408.
Carniglia, C. K. 1979. "Scalar Scattering Theory for Multilayer Optical Coatings." *Optical Engineering* 18: 104–115.
Chen, J., Q. Wang, and H. Li. 2010. "Microstructured Design for Light Trapping in Thin-Film Silicon Solar Cells." *Optical Engineering* 49 (8) (August). doi:10.1117/1.3476334.
Choy, T. C. 1999. *Effective Medium Theory: Principles and Applications*. Oxford University Press.

Cotter, J. E. 2005. "RaySim 6.0: A Free Geometrical Ray Tracing Program for Silicon Solar Cells." In *Conference Record of the Thirty-First IEEE Photovoltaic Specialists Conference, 2005*, 1165–1168. doi:10.1109/PVSC.2005.1488345.

Dewan, R., I. Vasilev, V. Jovanov, and D. Knipp. 2011. "Optical Enhancement and Losses of Pyramid Textured Thin-Film Silicon Solar Cells." *Journal of Applied Physics* 110 (1) (July 1). doi:10.1063/1.3602092.

Elsherbeni, A., and V. Demir. 2009. *The Finite-Difference Time-Domain Method for Electromagnetics with MATLAB® Simulations*. SciTech Publishing.

Elson, J. M., and J. M. Bennett. 1979. "Vector Scattering Theory." *Optical Engineering* 18 (April): 116–124.

Elson, J. M., J. P. Rahn, and J. M. Bennett. 1980. "Light Scattering from Multilayer Optics: Comparison of Theory and Experiment." *Applied Optics* 19 (5) (March 1): 669–679. doi:10.1364/AO.19.000669.

Green, M. A. 1981. *Solar Cells: Operating Principles, Technology, and System Applications*. Prentice Hall.

Green, M. A., K. Emery, Y. Hishikawa, W. Warta, and E. D. Dunlop. 2012. "Solar Cell Efficiency Tables (version 39)." *Progress in Photovoltaics: Research and Applications* 20 (1) (January 1): 12–20. doi:10.1002/pip.2163.

Guha, S., and J. Yang. 2010. "Thin Film Silicon Photovoltaic Technology—From Innovation to Commercialization." *MRS Online Proceedings* 1245. doi:10.1557/PROC-1245-A01-01.

Haase, C., and H. Stiebig. 2006. "Optical Properties of Thin-Film Silicon Solar Cells with Grating Couplers." *Progress in Photovoltaics: Research and Applications* 14 (7): 629–641. doi:10.1002/pip.694.

Haase, C., and H. Stiebig. 2007. "Thin-Film Silicon Solar Cells with Efficient Periodic Light Trapping Texture." *Applied Physics Letters* 91: 061116. doi:10.1063/1.2768882.

IEC 60904-3:2008. "IEC 60904-3:2008—Photovoltaic Devices—Part 3: Measurement Principles for Terrestrial Photovoltaic (PV) Solar Devices with Reference Spectral Irradiance Data."

Ingle, J. D., and S. R. Crouch. 1988. *Spectrochemical Analysis*. Prentice Hall.

Isabella, O., S. Solntsev, D. Caratelli, and M. Zeman. 2012. "3-D Optical Modeling of Thin-Film Silicon Solar Cells on Diffraction Gratings." *Progress in Photovoltaics: Research and Applications*. doi:10.1002/pip.1257.

Jäger-Waldau, A. 2011. *PV Status Report 2011*. JRC Scientific and Technical Reports.

Jin, J.-M. 2002. *The Finite Element Method in Electromagnetics*. 2nd ed. Wiley-IEEE Press.

Keller, A., S. Heinrich, and H. Niederreiter, eds. 2007. *Monte Carlo and Quasi-Monte Carlo Methods 2006*. Springer.

Krč, J., M. Zeman, O. Kluth, E. Smole, and M. Topič. 2003. "Effect of Surface Roughness of ZnO : Al Films on Light Scattering in Hydrogenated Amorphous Silicon Solar Cells RID A-5194-2008." *Thin Solid Films* 426 (1–2) (February 24): 296–304. doi:10.1016/S0040-6090(03)00006-3.

Krč, J., M. Zeman, F. Smole, and M. Topič. 2002. "Optical Modeling of a-Si:H Solar Cells Deposited on Textured Glass/SnO2 Substrates RID A-5194-2008." *Journal of Applied Physics* 92 (2) (July 15): 749–755. doi:10.1063/1.1487910.

Lacombe, J., O. Sergeev, K. Chakanga, K. von Maydell, and C. Agert. 2011. "Three Dimensional Optical Modeling of Amorphous Silicon Thin Film Solar Cells Using the Finite-Difference Time-Domain Method Including Real Randomly Surface Topographies." *Journal of Applied Physics* 110 (2) (July 15). doi:10.1063/1.3610516.

Lanz, T., B. Ruhstaller, C. Battaglia, and C. Ballif. 2011. "Extended Light Scattering Model Incorporating Coherence for Thin-Film Silicon Solar Cells RID B-2917-2010." *Journal of Applied Physics* 110 (3) (August 1). doi:10.1063/1.3622328.

Lipovšek, B., J. Krč, and M. Topič. 2011. Optical model for thin-film photovoltaic devices with large surface textures at the front side. *Inf. MIDEM*, 41,(4), 264–271.

Luque, A., and S. Hegedus, eds. 2011. *Handbook of Photovoltaic Science and Engineering.* 2nd ed. Wiley.

Myers, D. R., S. R. Kurtz, K. Emery, C. Whitaker, and T. Townsend. 2000. "Outdoor Meteorological Broadband and Spectral Conditions for Evaluating Photovoltaic Modules." In *Conference Record of the Twenty-Eighth IEEE Photovoltaic Specialists Conference, 2000,* 1202–1205..

Nevière, M., and E. Popov. 2002. *Light Propagation in Periodic Media: Differential Theory and Design.* CRC Press.

Oughstun, K. E., and N. A. Cartwright. 2003. "On the Lorentz-Lorenz Formula and the Lorentz Model of Dielectric Dispersion." *Optics Express* 11 (13) (June 30): 1541–1546.

Pflaum, C., and Z. Rahimi. 2011. "An Iterative Solver for the Finite-Difference Frequency-Domain (FDFD) Method for the Simulation of Materials with Negative Permittivity." *Numerical Linear Algebra with Applications* 18 (4) (August): 653–670. doi:10.1002/nla.746.

Photon International. 2002. "Market Survey on Cell Testers and Sorters." *The Photovoltaic Magazine* (10): 48–57.

Pieters, B. E, J. Krč, and M. Zeman. 2006. "Advanced Numerical Simulation Tool for Solar Cells—ASA5." In Proceedings of the IEEE 4th World Conference on Photovoltaic Energy Conversion, 1513–1516. doi:10.1109/WCPEC.2006.279758.

Saleh, B. E. A., and M. C. Teich. 2007. *Fundamentals of Photonics.* 2nd ed. Wiley-Interscience.

Schropp, R. E. I., and M. Zeman. 1998. *Amorphous and Microcrystalline Silicon Solar Cells: Modeling, Materials and Device Technology.* Springer.

Schulte, M., K. Bittkau, B. E. Pieters, S. Jorke, H. Stiebig, J. Huepkes, and U. Rau. 2011. "Ray Tracing for the Optics at Nano-textured ZnO-Air and ZnO-Silicon Interfaces RID H-3045-2011 RID G-2256-2011." *Progress in Photovoltaics* 19 (6) (September): 724–732. doi:10.1002/pip.1097.

Shah, A. V., ed. 2010. *Thin-Film Silicon Solar Cells.* EFPL Press.

Smith, F. G., T. A. King, and D. Wilkins. 2007. *Optics and Photonics: An Introduction.* 2nd ed. Wiley.

Solntsev, S., and M. Zeman. 2011. "Optical Modeling of Thin-Film Silicon Solar Cells with Submicron Periodic Gratings and Nonconformal Layers." *Energy Procedia* 10: 308–312. doi:10.1016/j.egypro.2011.10.196.

Springer, J., A. Poruba, and M. Vaneček. 2004. "Improved Three-Dimensional Optical Model for Thin-Film Silicon Solar Cells." *Journal of Applied Physics* 96 (9): 5329. doi:10.1063/1.1784555.

Stiebig, H., T. Brammer, T. Repmann, O. Kluth, N. Senoussaoui, A. Lambertz, and H. Wagner. 2000. "Light Scattering in Microcrystalline Silicon Thin Film Solar Cells." In *Proceedings of the Sixteenth European Photovoltaic Solar Energy Conference,* 549–552.

Street, R. A. 2000. *Technology and Applications of Amorphous Silicon.* Springer.

Synopsys. http://www.synopsys.com/Tools/OpticalDesign/Pages/default.aspx.

Taflove, A., and S. C. Hagness. 2005. *Computational Electrodynamics: The Finite-Difference Time-Domain Method.* 3rd ed. Artech House.

Troparevsky, M. C., A. S. Sabau, A. R. Lupini, and Z. Zhang. 2010. "Transfer-Matrix Formalism for the Calculation of Optical Response in Multilayer Systems: From Coherent to Incoherent Interference RID B-9571-2008." *Optics Express* 18 (24) (November 22): 24715–24721. doi:10.1364/OE.18.024715.

Tsang, L., J. A. Kong, and K.-H. Ding. 2000. *Scattering of Electromagnetic Waves: Theories and Applications*. Wiley-Interscience.

Wriedt, T., and W. Hergert, eds. 2012. *The Mie Theory: Basics and Applications*. Springer.

Zeman, M. 2012. Thin-Film Silicon Photovoltaics, Keynote Speech (3CP.1.1). 27th European Photovoltaic Solar Energy Conference and Exhibition, September 24–28, Frankfurt, Germany.

Zeman, M., J. A. Willemen, L. L. A. Vosteen, G. Tao, and J. W. Metselaar. 1997. "Computer Modelling of Current Matching in a-Si:H/a-Si:H Tandem Solar Cells on Textured TCO Substrates." *Solar Energy Materials and Solar Cells* 46 (2) (May): 81–99. doi:10.1016/S0927-0248(96)00094-3.

Zeman, M., R. A. C. M. M. van Swaaij, J. W. Metselaar, and R. E. I. Schropp. 2000. "Optical Modeling of a-Si:H Solar Cells with Rough Interfaces: Effect of Back Contact and Interface Roughness." *Journal of Applied Physics* 88 (11): 6436. doi:10.1063/1.1324690.

Zhao, L., Y. H. Zuo, C. L. Zhou, H. L. Li, H. W. Diao, and W. J. Wang. 2010. "A Highly Efficient Light-Trapping Structure for Thin-Film Silicon Solar Cells." *Solar Energy* 84 (1) (January): 110–115. doi:10.1016/j.solener.2009.10.014.

# One-Dimensional Semi-Coherent Optical Modeling

# 2

CONTENTS

## 2.1   INTRODUCTION

In this chapter we present an approach to one-dimensional (1-D) optical modeling of thin-film photovoltaic (PV) devices, which has turned out to be a successful way to describe (approximate) many of the optical effects in thin-film devices, such as coherent light propagation and scattering at nano-textured interfaces in multilayer structures. Based on the optical model described here, the simulator Sun*Shine* was developed and its licenses are used in various thin-film PV companies around the world. The model was applied to thin-film silicon, chalcopyrite, dye-sensitized, and organic solar cells. In this chapter we present the background, explain the details of the approach, and describe its implementation into a model and simulator. Let us explain what the adjective *semi-coherent* (modeling) refers to: in the approach, the part of direct (specular) light that does not get scattered when passing the nano-textured interfaces preserves its coherent nature, whereas for scattered light incoherent nature is applied. Therefore, the model is semi-coherent—coherent only for specular light.

## 2.2   BACKGROUND

In thin-film PV devices, lateral dimensions are usually much larger than the vertical one (thickness); therefore, 1-D modeling can be applied in many cases. Because the thicknesses of layers inside the devices are in the range of light wavelength (tens or hundreds of nanometers), coherent propagation of specular light within the thin layers has to be taken into account. *Coherent* means that the light is analyzed in terms of electromagnetic waves, which can interfere with each other, leading to pronounced interference fringes observed in spectral characteristics of the devices (reflectance, quantum efficiency). Besides specular light, scattered (diffused) light—as a consequence of the light-scattering process usually at randomly nano-textured interfaces, or due to the diffused component of incident illumination—is present in the structure. For scattered light, coherency effects are usually not observed (canceled) (Schropp and Zeman 1998). Therefore, we consider that in the structure, scattered light can be analyzed in an incoherent way, in terms of propagating light rays (photon fluxes) with certain intensity and direction. The term *intensity* refers to power density in mW/cm$^2$. Despite scattering at nano-textured interfaces, a part of light still remains specular and maintains its coherent nature as indicated by the moderate pronounced interference fringes in spectral characteristics of the

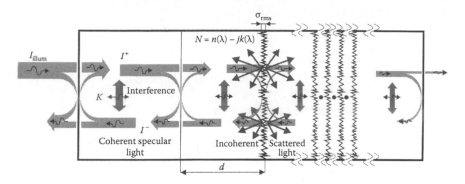

**FIGURE 2.1** Schematic of the optical situation in a multilayer thin-film solar cell structure with perpendicular specular illumination. Coherent propagation of specular light is indicated with wavy arrows, whereas the incoherent propagation of scattered light is denoted with straight, diverging arrows.

devices with nano-textured interfaces, as shown in Chapter 1. The model needs then to consider coherent propagation of specular light and incoherent propagation of scattered light, resulting in a semi-coherent optical model. In Figure 2.1 we present a schematic view inside the structure, with flat and nano-textured interfaces as a starting point of the analysis. How does the light propagate in such a structure? Specular coherent and incoherent scattered light components are schematically drawn in the figure. We assign the interaction, which occurs between forward-going and backward-going coherent light waves ($I^+$ and $I^-$), with the interference part $K$.

In our approach, the specular coherent and scattered incoherent light inside and outside thin-film structures are analyzed separately, although their connection has to be considered all the time. In the model we assume some restrictions. First is that the specular electromagnetic waves are applied and propagate only perpendicularly to the plan-parallel interfaces of the multilayer structure, in forward and backward directions as depicted in Figure 2.1. However, diffused incident illumination is allowed, but it is taken as incoherent light. At nano-textured interfaces, a part of the light is scattered each time when impinging the interface, presenting the origin of the scattered light that is incoherent (straight arrows in Figure 2.1). In the multilayer structure, each layer is described by the complex refractive index $N(\lambda)$ and its thickness $d$. One of the characteristic parameters of the texturization of a nano-textured interface is the vertical root-mean-square roughness, $\sigma_{rms}$ (see Section 3.2.5), also denoted in the figure.

The analysis of optical situation in thin-film structures can be divided into

- propagation of light throughout the layers and
- optical situation at the interfaces.

In the following sections, we explain the details of the analysis.

## 2.3  PROPAGATION OF LIGHT IN THE MODEL

We consider three different cases:

- propagation of specular coherent light in a thin layer,
- propagation of specular coherent light in a thick layer, and
- propagation of scattered incoherent light in a thin or a thick layer.

Coherency effects of specular light that are observed in measured spectral characteristics of the devices are related to coherent propagation of light in thin layers (thickness in the range of light wavelength). In the layers for which thickness exceeds the incoherency thickness $d_{incoh}$ [Equation (2.1)], coherency effects, in the form of dense interference fringes in the characteristics of the devices, are not observed anymore. These layers, whose thickness $d$ exceeds $d_{incoh}$, are assigned to thick layers.

$$d_{incoh} \cong \frac{\lambda^2}{2 \cdot \Delta\lambda \cdot n} \qquad (2.1)$$

In Equation (2.1), symbol $\lambda$ is the light wavelength (in free space), $\Delta\lambda_0$ is the bandwidth of the monochromator system that is used to determine spectral characteristics of thin-film structure, and $n$ is the real refractive index of the (thick) layer (Ley 1984). For example, considering a glass layer, which is used as a superstrate in thin-film PV devices, with its refractive index around 1.5 and a monochromator with $\Delta\lambda = 4$ nm and, for instance, selected wavelength $\lambda = 550$ nm, the thickness where coherent effects are not observed anymore is $d_{incoh} = 20.8$ μm. Typical thickness of glass superstrate is 0.6 – 3 mm; thus, it is a thick layer that has to be analyzed incoherently also for specular light. This means that the propagation of specular light has to be treated differently in thin and in thick layers. For the propagation of incoherent scattered light, the thickness of a layer does not play a role since it is analyzed incoherently in any case.

In solar illumination we consider that the coherency length of solar illumination $L_c$ is around 1 μm; thus, the layers whose thickness exceeds $L_c/n$ ($n$ is the refractive index of the layer) should be treated incoherently, as thick layers. However, under solar illumination we do not measure spectral characteristics of the devices; rather, we measure integrated parameters like short-circuit current density where information on coherency is not directly recognizable.

### 2.3.1  SPECULAR LIGHT IN A THIN LAYER

As mentioned, propagation of specular light in a thin layer is analyzed coherently, in terms of electromagnetic waves. The case of perpendicular propagation of specular light to the interfaces of the structure will be analyzed. Electric field

**Thin Layer**

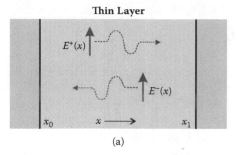

(a)

**Thin Layer**

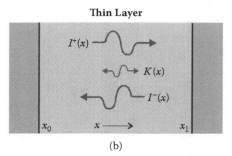

(b)

**FIGURE 2.2**    Propagation of forward-going (+) and backward-going (–) specular wave in a thin layer: (a) electric field strength and (b) light intensity representation.

strengths of an electromagnetic wave propagating in a forward direction (direction of "$x$" axis), $E^+$, and in a backward direction, $E^-$, are defined in Equations (2.2) and (2.3), respectively. Schematic representation of the propagation of light, perpendicular to the interfaces, is illustrated in Figure 2.2a.

$$E^+(x) = E^+(x_0) \cdot e^{-j\frac{2\pi N}{\lambda}(x-x_0)} \tag{2.2}$$

$$E^-(x) = E^-(x_0) \cdot e^{-j\frac{2\pi N}{\lambda}(x-x_1)} \tag{2.3}$$

The symbols $E^+(x_0)$ and $E^-(x_1)$ correspond to the electric field at the left and right interfaces (borders) of the layer, respectively, $N$ is the complex refractive index of the layer, $\lambda$ is the light wavelength (in free space), and $x$ is the spatial position in the layer. The equations determine the changes in the phase of the complex value of the electric field strength (determined by the real part of $N$, i.e., by the refractive index $n$) and attenuation of the amplitude (determined by imaginary part of $N$, i.e., by the extinction coefficient $k$) of the propagating wave. To get the common value of the electric field, $E(x)$, at a certain point (plane) in the layer, a superposition of forward and backward waves has to be applied, $E(x) = E^+(x) + E^-(x)$.

Once electric field strengths have been determined, we apply Equation (1.3), which is based on the Poynting vector, to obtain light intensities (Kong 1990). Vectors $E$ and $H$ can be substituted with scalars in Equation (1.3), since specular light propagates only in a perpendicular direction to the interfaces in our model, and thus, both electric and magnetic field strengths lie in the interface plane. They are also perpendicular to each other (see Figure 1.1b). The scalar of electric field strength is the sum $E = E^+ + E^-$. The scalar of magnetic field strength can be expressed by electric field strengths as $H = Y_0 N (E^+ - E^-)$, where $Y_0$ is the admittance of free space ($Y_0 = 2.6544 \cdot 10^{-3}$ S) and $N$ is the refractive index of the material. The negative sign in front of $E^-$ reflects the opposite direction of $H^-$ with respect to $H^+$, considering the opposite direction of backward-going light. Please note that in Equation (1.3) the conjugate value of $H$ has to be considered and that vector product is substituted by scalar product. When we multiply the $E$ and $H^*$ components, we come to the terms given in Equation (2.4), representing light intensities. We see that besides forward-going light intensity, $I^+$, and backward-going light intensity, $I^-$, there exists also an interference part $K^-$ in the case of absorbing medium ($k \neq 0$) as sketched in Figure 2.2b. The $K^-$ can have at a certain point $x$ forward or backward direction, depending on the sign of the term $\text{Im}[E^- \cdot E^{+*}]$.

$$I = \frac{1}{2} Y_0 \cdot n \cdot |E^+|^2 - \frac{1}{2} Y_0 \cdot n \cdot |E^-|^2 - Y_0 \cdot k \cdot \text{Im}[E^- \cdot E^{+*}]$$

$$= I^+ - I^- - K \tag{2.4}$$

When including specular light that incidents non-perpendicularly to the interfaces of the structure, vectors have to be considered in Equations (2.2) and (2.4), including information about the actual direction of light. Often, transfer matrix formalism is used in such cases to resolve optical situation in a more simple way (Troparevsky et al. 2010).

### 2.3.2   SPECULAR LIGHT IN A THICK LAYER

As mentioned at the beginning of Section 2.3, special attention has to be paid when simulating the propagation of coherent light in a thick layer with the thickness $d \geq d_{\text{incoh}}$. We want to avoid dense interference fringes in spectral characteristics (they would be present in an ideal case but are not observed in measured characteristics) (Stiebig et al. 1994). To demonstrate this effect, we show in Figure 2.3 an example of simulation of total reflectance from an amorphous silicon single-junction solar cell in which perpendicular propagation of specular light in a front glass superstrate is taken coherently (dashed line) and incoherently (full line). For the latter case a good agreement with the measured characteristics could be found, whereas the first simulation exhibits unrealistic interference fringes.

Different approaches of incoherent treatment of a thick layer surrounded by thin coherent layers (on one or both sides) in a multilayer structure have been

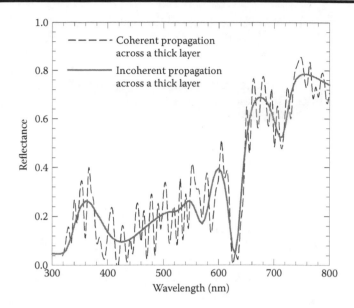

**FIGURE 2.3**    Simulated total reflectance from a thin-film a-Si:H solar cell structure with a thick glass superstrate ($d_{glass}$ = 1.5 mm) considering coherent and incoherent propagation of light throughout the thick superstrate. The wavelength step in both simulations was 10 nm.

reported (Santbergen 2008; Troparevsky et al. 2010; Jung et al. 2011). Here we describe an approach based on thick-layer thickness averaging (Stiebig 1997). This approach enables us to treat the propagation of specular light in a thick layer as coherent, however, averaging the results over four simulations, where the thickness of the thick layer has to be changed. The thickness has to be modified according to the following rule given in Equation (2.5) (Stiebig 1997).

$$d_1 = d_{thick} - \frac{\lambda}{4} \cdot \frac{1.5}{n}$$

$$d_2 = d_{thick}$$

$$d_3 = d_{thick} + \frac{\lambda}{4} \cdot \frac{1.5}{n} \qquad (2.5)$$

$$d_4 = d_{thick} + \frac{\lambda}{2} \cdot \frac{1.5}{n}$$

In Equation (2.5), $d_{thick}$ is the actual thickness of the thick layer, $n$ is the refractive index (real part of $N$) of the layer, and $\lambda$ is the light wavelength in free space. With such modification of thicknesses, the interference fringes in simulations, originating from coherent effects in a thick layer, can be cancelled out.

The interferences related to thin layers in the multilayer structure still remain and reassemble the ones observed in the measurements (see Figure 2.3, where the incoherency of the thick glass substrate was obtained with the approach described). Another possible way of eliminating interference fringes would be averaging of simulation results obtained for certain intervals of neighboring wavelengths. In this case more than four simulations would be required, according to our experience.

### 2.3.3   SCATTERED LIGHT IN A THIN OR THICK LAYER

Scattered light rays treated as incoherent light are described by their intensities $I$ (power densities) and directions of propagations. The direction of a scattered ray can be described by the corresponding angle $\varphi$ defined toward the $x$-axis (Figure 2.4). The optical path of a scattered ray, propagating outside the specular direction, is prolonged by a factor of $1/\cos\varphi$. Intensities of scattered rays propagating in forward and backward direction, $I^+(\varphi,x)$ and $I^-(\varphi,x)$, are defined by Equations (2.6) and (2.7), respectively. The equations are applicable for a thin or thick layer.

$$I^+(\varphi,x)=I^+(\varphi,x_0)\cdot e^{-\frac{4\pi k}{\lambda}\cdot\frac{(x-x_0)}{\cos\varphi}}=I^+(\varphi,x_0)\cdot e^{-\frac{4\pi\cdot\frac{k}{\cos\varphi}}{\lambda}\cdot(x-x_0)} \qquad (2.6)$$

$$I^-(\varphi,x)=I^-(\varphi,x_1)\cdot e^{-\frac{4\pi k}{\lambda}\cdot\frac{(x-x_1)}{\cos\varphi}}=I^-(\varphi,x_1)\cdot e^{-\frac{4\pi\cdot\frac{k}{\cos\varphi}}{\lambda}\cdot(x-x_1)} \qquad (2.7)$$

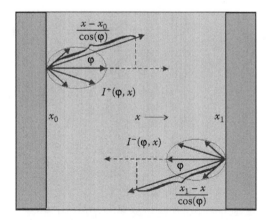

**FIGURE 2.4**   Propagation of scattered light rays in a thin or thick layer (in the figure we assume the origin of light scattering at both interfaces).

The symbols $I^+(\varphi, x_0)$ and $I^-(\varphi, x_1)$ are the intensities of the forward and backward propagating rays defined at the interfaces (borders) of the layer. The common intensity in a specific direction and location in the layer can be determined in this case as $I(\varphi, x) = I^+(\varphi, x) - I^-(\varphi, x)$.

The prolongation factor $1/\cos\varphi$ can also be applied to the extinction coefficient $k$ [last term of Equations (2.6) and (2.7)]. In this case, the propagation of the ray can be projected to the specular direction ($\varphi = 0$), whereas the increased absorption due to a prolonged optical path is transferred to increased extinction coefficient $k$.

## 2.4 OPTICAL SITUATION AT THE INTERFACES

Besides propagation of light throughout individual layers, we have to define optical situations at the interfaces as well. Two types of interfaces will be considered here: flat and randomly nano-textured. In general, we also have periodically textured interfaces that have been introduced in thin-film photovoltaic devices; these are discussed in Chapters 4 and 6. First, we analyze the incidence of specular coherent and scattered incoherent light at a flat interface.

### 2.4.1 SPECULAR LIGHT AT A FLAT INTERFACE

Let us consider perpendicular incidence of specular coherent light at a flat interface. The reflectance and the transmittance of electromagnetic waves, described by electric field strength $E$, at a flat interface are determined by Fresnel's coefficients $r_0$ and $t_0$ (Kong 1990). By considering that the light falls at the interface from the left (l) side (layer or medium), the reflectance, $r_{0l}$, and transmittance (from the left to right side "lr" of the interface), $t_{0lr}$, are defined by Equations (2.8) and (2.9). The situation is schematically represented in Figure 2.5.

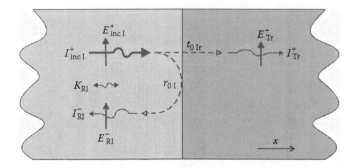

**FIGURE 2.5**  Optical situation at a flat interface. A coherent specular light ($E^+_{\text{inc l}}$) is applied from the left side of the interface only, due to simplicity.

The symbols $I^+_{\text{inc l}}$, $I^-_{\text{R l}}$, $K_{\text{R l}}$, and $I^+_{\text{T r}}$ represent the light intensities that can be calculated from the corresponding electric field strengths.

$$r_{01} = \frac{N_1 - N_r}{N_1 + N_r} \tag{2.8}$$

$$t_{0\,\text{lr}} = \frac{2N_1}{N_1 + N_r} \tag{2.9}$$

$$t_{0\,\text{lr}} = 1 + r_{01} \tag{2.10}$$

The symbols $N_1$ and $N_r$ correspond to the complex refractive indexes of the left and the right layer, respectively. The wavelength dependency of complex refractive indexes, and consequently of Fresnel's coefficients, is not indicated in the equations. The direct relation between $r_{01}$ and $t_{0\,\text{lr}}$ can be written as given in Equation (2.10).

The reflected electric field component at the left side, $E^-_{\text{R l}}$, and the transmitted one on the right side of the interface, $E^+_{\text{T r}}$, can be determined by Equations (2.11) and (2.12), where $E^+_{\text{inc l}}$ represents the electric field strength of the incident wave. According to the direction of propagation of the waves with respect to the $x$-axis, "+" and "−" superscripts are added to the symbols.

$$E^-_{\text{R l}} = r_{01} \cdot E^+_{\text{inc l}} \tag{2.11}$$

$$E^+_{\text{T r}} = t_{0\,\text{lr}} \cdot E^+_{\text{inc l}} \tag{2.12}$$

Because of the presence of forward- and backward-going waves in multilayer thin-film structures, the light impinges the interface from both the left and the right side. Thus, in general we have to consider two incident components, $E^+_{\text{inc l}}$ from the left and $E^-_{\text{inc r}}$ from the right side. This situation is presented in Figure 2.6. Reflectance for the wave approaching from the right side, $r_{0\,\text{r}}$, and the corresponding transmittance $t_{0\,\text{rl}}$ are defined similarly as their counterparts $r_{01}$ and $t_{0\,\text{lr}}$, considering the exchange of the index "l" with "r" and vice versa in Equations (2.8) and (2.9). On the left side, there exists one forward-going wave, $E^+_{\text{inc l}}$, and two backward-going waves, the reflected one $E^-_{\text{R l}}$ and the transmitted one $E^-_{\text{T r}}$, which can be combined into one wave with common electric field strength $E^-_1 = E^-_{\text{R l}} + E^-_{\text{T r}}$. On the right side the two forward-going waves can be combined into one, $E^+_r = E^+_{\text{R r}} + E^+_{\text{T r}}$.

In one approach to calculating specular light distribution in a multilayer structure, which we also follow in the optical simulator Sun*Shine*, the calculation starts at the rear (most right side) of the structure (this approach is explained in Section 2.6). This means that at an interface, the electric field strengths at the right side, $E^-_{\text{inc r}}$ and $E^+_r$, are predefined and the components at the left side of the interface, $E^+_{\text{inc l}}$ and $E^-_l$, are unknowns. From Equations

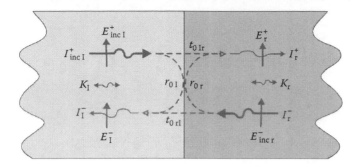

**FIGURE 2.6** Optical situation at a flat interface. A coherent specular light is applied from both the left ($E^+_{inc\,l}$) and the right ($E^-_{inc\,r}$) side of the interface at the same time.

(2.11) and (2.12), and by considering the superposition rule for the components propagating in the same direction, the equations to calculate $E^+_{inc\,l}$ and $E^-_l$ can be derived.

$$E^+_{inc\,l} = \frac{1}{t_{0\,lr}} \cdot E^+_r - \frac{r_{0r}}{t_{0\,lr}} \cdot E^-_{inc\,r} \tag{2.13}$$

$$E^-_l = \frac{r_{0l}}{t_{0\,lr}} \cdot E^+_r - \frac{r_{0r} \cdot r_{0l} - t_{0\,rl} \cdot t_{0\,lr}}{t_{0\,lr}} \cdot E^-_{inc\,r} \tag{2.14}$$

By substituting $r_{0\,l}$, $r_{0\,r}$, $t_{0\,lr}$, and $t_{0\,rl}$ with the corresponding expressions, including complex refractive indexes $N_r$ and $N_l$, the $E^+_{inc\,l}$ and $E^-_l$ can be written as given in Equations (2.15) and (2.16).

$$E^+_{inc\,l} = \frac{N_l + N_r}{2N_l} E^+_r + \frac{N_l - N_r}{2N_l} E^-_{inc\,r} \tag{2.15}$$

$$E^-_l = \frac{N_l - N_r}{2N_l} E^+_r + \frac{N_l + N_r}{2N_l} E^-_{inc\,r} \tag{2.16}$$

After $E^-_{inc\,r}$, $E^+_r$, and $E^+_{inc\,l}$, $E^-_l$ are defined, the corresponding light intensities at the left, $I^+_l - I^-_l - K_l = I_l$, and right side of the interface, $I^+_r - I^-_r - K_r = I_r$, can be determined by determining the real part of the Poynting vector [Equation (2.4)]. At the interface (and at each point in the structure) the common light intensity $I = I^+ - I^- - K$ has to be continuous along the propagation axis in order to fulfill the energy conservation law. This holds also for the common light intensities at the left and right sides of the interface. Given the previously presented equations, it can be shown that $I_l = I_r$ as required by the conservation law. The same condition is fulfilled also for situations analyzed in Sections 2.4.2, 2.5.1, and 2.5.2.

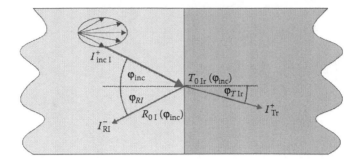

**FIGURE 2.7**    Optical situation at a flat interface. A selected ray of scattered incoherent light, $I^+_{inc\,l}$, is impinging the interface from the left side under incident angle $\varphi_{inc}$.

In the case of non-perpendicular incidence of specular light, one has to consider also the polarization of light (direction of the vectors of electric and magnetic field strengths) and the incident angle, resulting in more complex equations.

### 2.4.2    SCATTERED LIGHT AT A FLAT INTERFACE

Scattered light rays impinge the interfaces under different incident angles. Incoherent rays are described by their intensities $I$ and the directions (angles $\varphi$). By knowing the intensity and the angle of propagation of the incident ray, $I^+_{inc}$ and $\varphi_{inc}$, the corresponding reflected, $I_R$, and transmitted, $I_T$, components and their angles of propagation, $\varphi_R$ and $\varphi_T$, have to be determined at an interface. The situation is illustrated in Figure 2.7 for the case of a selected non-perpendicular incident ray of scattered light, falling on the interface from the left (l) side. The corresponding subscripts "l" and "lr" are added to the symbols using the same methodology as in the previous section.

First the reflectance $R_{0\,l}$ and the transmittance $T_{0\,lr}$, referring to light intensities and not to electric field strengths in this case, are defined. $R_{0\,l}$ and $T_{0\,lr}$ depend on the angle of the incident ray, $\varphi_{inc}$, and its polarization (combination of horizontal and vertical polarization*). The total reflectance $R_{0\,l}(\varphi_{inc})$ can be constructed from the contribution of horizontally (H) and of vertically (V) polarized light that are present in the incident ray [Equation (2.17)].

$$R_{0\,l}(\varphi_{inc}) = m \cdot R_{H\,l}(\varphi_{inc}) + (1-m) \cdot R_{V\,l}(\varphi_{inc}) \qquad 0 < m < 1 \qquad (2.17)$$

The parameter $m$ defines the ratio between the two polarizations. Since the polarization of incoherent scattered rays is most commonly not traced, the

---

\* There are different names used for polarization of light. *Horizontal polarization* refers to transverse electric (TE) or s-polarization. *Vertical polarization* refers to transverse magnetic (TM) or p-polarization.

parameter $m$ is usually set to 0.5, meaning half horizontal and half vertical (i.e., circular) polarization. Further on, the reflectances $R_{H1}(\varphi_{inc})$ and $R_{V1}(\varphi_{inc})$ are

$$R_{H1}(\varphi_{inc}) = \left| \Gamma_{H1}(\varphi_{inc}) \right|^2 \tag{2.18}$$

$$R_{V1}(\varphi_{inc}) = \left| \Gamma_{V1}(\varphi_{inc}) \right|^2 \tag{2.19}$$

where $\Gamma_{H1}(\varphi_{inc})$ is the reflectance for electric field strength of horizontally polarized light, whereas $\Gamma_{V1}(\varphi_{inc})$ is the reflectance of magnetic field strength of vertically polarized light and are determined by Equations (2.20) and (2.21) (Kong 1990).

$$\Gamma_{H1}(\varphi_{inc}) = \frac{\cos\varphi_{inc} - \sqrt{\frac{N_r^2}{N_l^2} - \sin^2\varphi_{inc}}}{\cos\varphi_{inc} + \sqrt{\frac{N_r^2}{N_l^2} - \sin^2\varphi_{inc}}} \tag{2.20}$$

$$\Gamma_{V1}(\varphi_{inc}) = \frac{\sqrt{\frac{N_r^2}{N_l^2} - \sin^2\varphi_{inc}} - \frac{N_r^2}{N_l^2}\cdot\cos\varphi_{inc}}{\sqrt{\frac{N_r^2}{N_l^2} - \sin^2\varphi_{inc}} + \frac{N_r^2}{N_l^2}\cdot\cos\varphi_{inc}} \tag{2.21}$$

Symbols $N_l$ and $N_r$ are complex refractive indexes of the left and right layer again. For a normal incidence of light ray ($\varphi_{inc} = 0$), it can be shown that $\Gamma_H = \Gamma_V = r_0$.

To determine the transmittance, $T_{0\,lr}(\varphi_{inc})$, the energy conservation law, which applies to the situation at a non-absorbing interface as $R + T = 1$, is considered in this case, resulting in Equation (2.22).

$$T_{0\,lr}(\varphi_{inc}) = 1 - R_{01}(\varphi_{inc}) \tag{2.22}$$

The intensity of reflected and transmitted component now can be determined as

$$I_{R1}^- = R_{01}(\varphi_{inc})\cdot I_{inc1}^+ \tag{2.23}$$

$$I_{Tr}^+ = T_{0\,lr}(\varphi_{inc})\cdot I_{inc1}^+ \tag{2.24}$$

In addition to determining the intensities of reflected and transmitted light rays, their directions of propagation have to be defined. From the electromagnetic wave theory or ray optics, it is known that the angle of reflected beam, $\varphi_R$, is equal to the incident angle, $\varphi_{inc}$, as shown in Equation (2.25).

$$\varphi_{R1} = \varphi_{inc} \tag{2.25}$$

To determine the angle of transmitted light ray, $\varphi_{T\,lr}$, taking into consideration that the layers can be absorbing, the complex form of Snell's law has to be considered [Equation (2.26)],

$$\varphi_{T\,lr} = \text{Re}\left[\arcsin\left(\frac{N_l}{N_r}\sin\varphi_{inc}\right)\right] \tag{2.26}$$

where the "arcsin" function of a complex number $(N_l/N_r \cdot \sin\varphi_{inc}) = z$ can be defined as

$$\arcsin(z) = i\ln(iz + \sqrt{1-z^2}) \tag{2.27}$$

and the natural logarithm of a complex number $iz + \sqrt{1-z^2} = w$ can be defined as

$$\ln(w) = \ln(|w|) + i[\arg(w)]. \tag{2.28}$$

In multilayer structures, the scattered rays are approaching the interfaces from both the left and the right side. The right-side illumination case can be analyzed in the same way as the described left-side illumination case, by considering an appropriate exchange of indexes "l" and "r." Here, the combined case of left- and right-side illumination is not shown as it was for the specular coherent illumination because in the presented modeling approach, incoherent illumination from left and right sides of the interface are analyzed separately, and then the superposition of light intensities is considered.

## 2.5   DESCRIPTION OF LIGHT SCATTERING AT A NANO-TEXTURED INTERFACE

Thus far, we have described the optical situation at a flat interface. In this section we deal with the situation at a textured interface.

We know that at a textured interface with the surface roughness in the range of nanometers a part of incident light (coherent or incoherent) is scattered, in reflection and in transmission, whereas the rest remains specular, as schematically shown in Figure 2.8. The scattering process depends on interface morphology (texturization), complex refractive indexes of the incident medium and the medium in transmission, and on the type of incident light. It is most typical that random texturization is present in thin-film solar cells. Vertical and lateral dimensions of the texturization features are in the range of light wavelength, and the features can have sharp edges (see example of pyramid-like texture in Figure 2.9). All of these factors lead to a rather complex

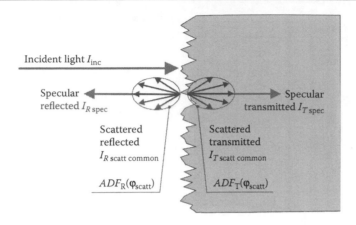

**FIGURE 2.8**  Schematic representation of light scattering at a nano-textured surface. Specular and scattered (diffused) components are present in reflection and in transmission.

optical situation at a textured interface. The relationship between surface morphology and light-scattering properties becomes complicated and analytically or numerically difficult to solve. In 1-D modeling approaches, approximations are used to define light-scattering properties of nano-textured interfaces (see Section 3.2.6).

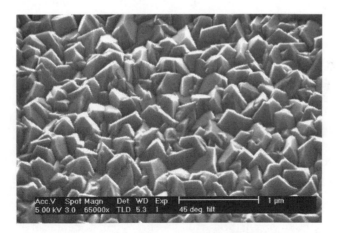

**FIGURE 2.9**  An example of surface nano-texture used in thin-film silicon solar cells. Scanning electron microscopy (SEM) picture of the surface of Asahi U-type SnO$_2$:F transparent conductive oxide (TCO) on glass superstrate. The vertical root-mean-square roughness, $\sigma_{rms}$, of the surface texturization is 40 nm.

To describe the amount and properties of scattered light at a textured interface, so-called descriptive scattering parameters can be used. There are two of them:

- haze ($H$) and
- angular distribution function ($ADF$) of scattered light.

Both parameters are wavelength dependent in general.

The haze parameter determines the level of light scattering and is defined as the ratio between the common scattered, $I_{\text{scatt common}}$, and the total, $I_{\text{tot}}$ (specular $I_{\text{spec}}$ + scattered), light intensity for the case of reflected ($H_R$) and transmitted light ($H_T$) at a textured interface, as given in Equations (2.29) and (2.30).

$$H_R = \frac{I_{\text{R scatt common}}}{I_{\text{R spec}} + I_{\text{R scatt common}}} = \frac{I_{\text{R scatt common}}}{I_{\text{R tot}}} = \frac{R_{\text{dif}}}{R_{\text{tot}}} \tag{2.29}$$

$$H_T = \frac{I_{\text{T scatt common}}}{I_{\text{T spec}} + I_{\text{T scatt common}}} = \frac{I_{\text{T scatt common}}}{I_{\text{T tot}}} = \frac{T_{\text{dif}}}{T_{\text{tot}}} \tag{2.30}$$

Instead of the ratio between light intensities, the ratio between diffused (scattered) and total reflectance ($R_{\text{dif}}/R_{\text{tot}}$) and transmittance ($T_{\text{dif}}/T_{\text{tot}}$) can be used to define $H_R$ and $H_T$, respectively. Methods of analytical and experimental determination of haze parameters are described in more detail in Section 3.2.6. The haze parameters of nano-textures are usually wavelength dependent. Their values can vary between 0 and 1, where 0 means no scattering (the case of a flat interface) and value 1 corresponds to full scattering (typically at very rough interface).

An example of measured $H_R$ and $H_T$ of the sample with surface texturization presented in Figure 2.9 is shown in Figure 2.10. As mentioned, details are discussed in Section 3.2.6.

The $ADF$ of scattered light describes the directional (angular) dependency of scattered light (how the light is scattered in different directions). It is defined for reflected, $ADF_R$, and transmitted light, $ADF_T$ (see schematic representation in Figure 2.8). Basically, $ADF$s are functions of scattering angles, $\varphi_{\text{scatt}}$, and usually also of the incident angle of the incoming light, $\varphi_{\text{inc}}$. In addition, they may also depend on light wavelength $\lambda$. Typically, $ADF$s are presented in their normalized form (values between 0 and 1, the sum of the components corresponding to different scattering angles is set to 1). Thus, the value of the $ADF$, corresponding to a certain angle, gives information about the relative share of the scattered light intensity that propagates in a certain direction, with respect to the common intensity of scattered light in reflection or in transmission. Considering this, the $ADF_R$ and $ADF_T$ can be written as given in Equations (2.31) and (2.32),

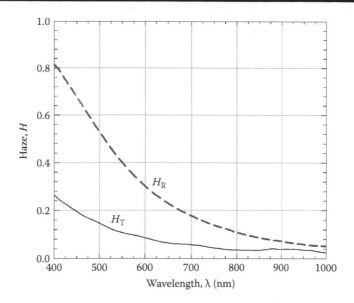

**FIGURE 2.10** An example of measured haze parameters (in surrounding air) in reflection ($H_R$) and in transmission ($H_T$) of Asahi U-type TCO superstrate whose surface texture is shown in Figure 2.9. Higher haze is observed for shorter wavelengths where the ratio between the size of surface texture features and the light wavelength is higher than that at longer wavelengths.

where the dependence on $\varphi_{inc}$ and $\lambda$ is not indicated. More details about $ADFs$ of nano-textured interfaces are given in Section 3.2.6.

$$ADF_R(\varphi_{scatt}) = \frac{I_{R\ scatt}(\varphi_{scatt})}{I_{R\ scatt\ common}} \qquad (2.31)$$

$$ADF_T(\varphi_{scatt}) = \frac{I_{T\ scatt}(\varphi_{scatt})}{I_{T\ scatt\ common}} \qquad (2.32)$$

An example of measured $ADF_R$ and $ADF_T$ is shown in Figure 2.11. The measurements correspond to the surface morphology given in Figure 2.9.

### 2.5.1 SPECULAR LIGHT AT A TEXTURED INTERFACE

As mentioned in the previous section, at a nano-textured interface a part of reflected and transmitted light can remain specular. The specular components of light preserve their coherent nature; thus, they must be analyzed

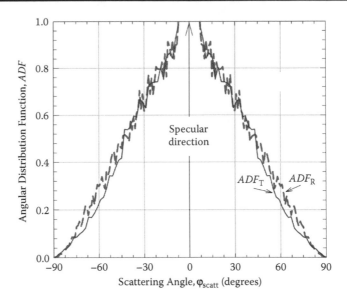

**FIGURE 2.11**    An example of measured angular distribution function (in surrounding air) in reflection ($ADF_R$) and in transmission ($ADF_T$) of Asahi U-type TCO superstrate. The measurements were performed at $\lambda = 633$ nm and $\varphi_{inc} = 0°$. A similar shape of $ADF_R$ and $ADF_T$ is observed.

as electromagnetic waves. The optical situation for a specular incident light approaching a textured interface from the left side is illustrated in Figure 2.12.

To determine reflected and transmitted specular components, specular reflectance, $r_{spec\,l}$, and specular transmittance, $t_{spec\,lr}$, for electric field strengths of the electromagnetic wave are defined in Equations (2.33) and (2.34)

$$r_{spec\,l} = r_{0\,text\,l} \cdot \sqrt{1 - H_{R\,l}} \qquad (2.33)$$

$$t_{spec\,lr} = t_{0\,text\,lr} \cdot \sqrt{1 - H_{T\,lr}} \qquad (2.34)$$

where $H_{R\,l}$ and $H_{T\,lr}$ are haze parameters of the interface for reflected and transmitted light and $r_{0\,text\,l}$ and $t_{0\,text\,lr}$ are total reflectance and transmittance for the electric field strength of the textured interface. Indexes "l" and "lr" correspond to the situation of left-side illumination. For large values of haze parameters, $r_{spec\,l}$ and $t_{spec\,lr}$ are small, resulting in smaller specular components of reflected and transmitted light and, consequently, larger scattered components, as shown later. On the contrary, for haze values equal to zero (e.g., in the case of a flat

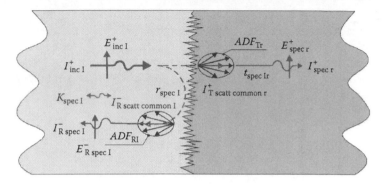

**FIGURE 2.12**  Optical situation at a textured interface. Coherent specular light $E^+_{\text{inc I}}$ is applied from the left side of the interface.

interface), $r_{\text{spec I}}$ and $t_{\text{spec Ir}}$ become equal to $r_{0\,\text{text I}}$ and $t_{0\,\text{text Ir}}$, respectively, which, in the case of a flat interface, are equal to Fresnel's coefficients $r_{01}$ and $t_{0\,\text{Ir}}$. However, for textured interfaces $r_{0\,1\,\text{text}}$ can be lower than $r_{01}$ (and $t_{0\,\text{text Ir}}$ consequently higher) due to antireflective effects caused by texturization (see Section 3.2.7). As a first approximation it can be taken that $r_{0\,\text{text I}} = r_{01}$ and $t_{0\,\text{text Ir}} = t_{0\,\text{Ir}}$. In any case, the relation between $r_{0\,\text{text I}}$ and $t_{0\,\text{text Ir}}$ based on Equation (2.10) (i.e., $t_{0\,\text{text Ir}} = 1 + r_{0\,\text{text I}}$) still holds, to fulfill the energy conservation law. Thus, by defining $r_{0\,\text{text I}}$, $t_{0\,\text{text Ir}}$ is defined. A similar approach to applying haze to electric field strength is taken in Springer, Poruba, and Vaneček (2004).

For the case of specular light, illuminations from both sides of the interface have to be considered at once because of the chosen method of calculation of the specular light (see Section 2.6). The situation is illustrated in Figure 2.13.

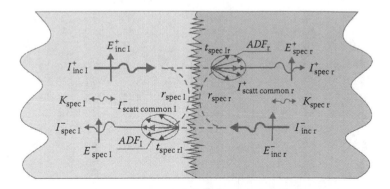

**FIGURE 2.13**  Optical situation at a textured interface. Coherent specular light is applied from both left ($E^+_{\text{inc I}}$) and right ($E^-_{\text{inc r}}$) sides of the interface.

Specular reflectance and transmittance of an interface have to be defined for the light approaching from the left ($r_{\text{spec l}}$, $t_{\text{spec lr}}$) and right sides of the interface ($r_{\text{spec r}}$, $t_{\text{spec rl}}$). For the later two, $r_{0\,\text{text r}}$, $t_{0\,\text{text rl}}$, $H_{\text{R l}}$, $H_{\text{T lr}}$, $H_{\text{R r}}$, and $H_{\text{T rl}}$ have to be considered. Consistent refractive index exchange, corresponding to the left and right layers, have to be taken into account in the equations that define mentioned parameters.

The electric field strengths $E^+_{\text{inc l}}$ and $E^-_{\text{spec l}}$ can be calculated from precalculated (known) $E^+_{\text{spec r}}$ and $E^-_{\text{inc r}}$ by using Equtions (2.13) and (2.14) (Section 2.4.1) by applying specular reflectance and transmittance parameters corrected by haze instead of Fresnel's coefficients.

From known electric field strengths of specular components on both sides of the interface, we can determine the corresponding light intensities $I^+_{\text{inc l}}$, $I^-_{\text{spec l}}$, $K_{\text{spec l}}$, $I^+_{\text{spec r}}$, $I^-_{\text{spec r}}$, $K_{\text{spec r}}$ from the Poynting vector as in the case of specular light at a flat interface (Section 2.4.1).

Thus far, determination of specular components at a textured interface has been described. For determination of scattered components on the left and right side of the interface, energy conservation law can be used, resulting in Equation (2.35) (example for scattered component in the left layer, the same equation holds for the right layer, considering the exchange in index "l" with "r").

$$I^-_{\text{scatt common l}} = I_{\text{tot l}} - I_{\text{spec l}} \tag{2.35}$$

The quantity $I_{\text{scatt common l}}$ is the common scattered light intensity (sum of intensities propagating in all scattering angles, in this case in a backward direction; therefore, the denotation "−" is used). It consists of reflected scattered light due to the left-side illumination and transmitted scattered light due to the right-side illumination. The quantity $I_{\text{spec l}}$ represents specular components and can be calculated as $I_{\text{spec l}} = I^+_{\text{inc l}} - I^-_{\text{spec l}} - K_{\text{spec l}}$. The quantity $I_{\text{tot l}}$ represents the total intensity, that is, sum of specular and scattered components. In setting the haze parameters in Equations (2.33) and (2.34) to zero, $I_{\text{tot l}}$ is represented by only specular components, whereas the value of $I_{\text{tot l}}$ remains the same as in the case of non-zero haze parameters (since only the share of specular and scattered light is changed). Therefore, $I_{\text{tot l}}$ can be determined under these specular conditions. Then, $I_{\text{scatt common l}}$ can be calculated by Equation (2.35).

Angular distribution functions of scattered light components on the left and right sides of the interface have to be defined as well. Since illumination is applied from both sides, the scattered components comprise reflected and transmitted light. Therefore, the actual $ADF_l$ is a combination of the $ADF_{\text{R l}}$ and $ADF_{\text{T rl}}$. The share of each can be defined by the share of reflected and transmitted scattered light at the left side of the interface. Thus, the $ADF_l$ can be calculated as given in Equation (2.36). All $ADFs$ correspond to $\varphi_{inc} = 0$ (specular incident illumination) in this case.

$$ADF_1(\varphi_{scatt}) = \frac{I_{spec\,1}^{+'} \cdot R_{dif\,1} \cdot ADF_{R\,1}(\varphi_{scatt}) \;+\; I_{spec\,r}^{-'} \cdot T_{dif\,rl} \cdot ADF_{T\,rl}(\varphi_{scatt})}{I_{spec\,1}^{+'} \cdot R_{dif\,1} \;+\; I_{spec\,r}^{-'} \cdot T_{dif\,rl}} \qquad (2.36)$$

where

$$I_{1\,spec}^{+'} = \begin{cases} I_{spec\,1}^{+} - K_{spec\,1} & if \quad K_{spec\,1} < 0 \\ I_{spec\,1}^{+} & if \quad K_{spec\,1} \geq 0 \end{cases} \qquad (2.37)$$

$$I_{r\,spec}^{-'} = \begin{cases} I_{spec\,r}^{-} & if \quad K_{spec\,r} < 0 \\ I_{spec\,r}^{-} + K_{spec\,r} & if \quad K_{spec\,r} \geq 0 \end{cases} \qquad (2.38)$$

$$R_{dif\,1} = R_{0\,text\,1} - R_{spec\,1} = R_{0\,text\,1} - \left| r_{spec\,1} \right|^2 \qquad (2.39)$$

$$T_{dif\,rl} = T_{0\,text\,1} - T_{spec\,rl} = (1 - R_{0\,text\,r}) - (1 - \left| r_{spec\,r} \right|^2) \qquad (2.40)$$

and $ADF_{R\,1}(\varphi_{scatt})$ and $ADF_{T\,rl}(\varphi_{scatt})$ are input parameters.

When $I_{scatt\,common\,1}^{-}$ and $ADF_1(\varphi_{inc})$ are known, the intensity of a particular scattered ray at the left side of the interface can be calculated as

$$I_{scatt}^{-}(\varphi_{scatt}) = I_{scatt\,common\,1}^{-} \cdot ADF_1(\varphi_{scatt}). \qquad (2.41)$$

All of the scattered rays at the left side of the interface that are a consequence of light scattering at that interface are propagating away from the interface. Thus, on the left side of the interface they are moving in a backward direction, whereas on the right side of the interface they are propagating in a forward direction. The $ADF_r(\varphi_{scatt})$ can be determined by using the same equations as given previously for the $ADF_1(\varphi_{scatt})$, by considering the "l" and "r" index exchange.

## 2.5.2  SCATTERED LIGHT AT A TEXTURED INTERFACE

In the case of scattered light, incoherent rays impinge the interface under different directions in the presented model. At a nano-textured interface, a part of scattered light rays is re-scattered, whereas the rest remain specular in the sense that their propagation direction is the same as it would be in the case of

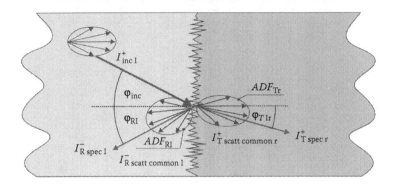

**FIGURE 2.14**   Optical situation at a textured interface. A ray of incoherent scattered light is applied from the left ($I^+_{inc\,l}$) side of the interface.

a flat interface (here *specular* does not also mean *coherent*). Because all of the rays are assumed to be incoherent here, calculation for each incident ray can be done independently, as in the case of a flat interface in Section 2.4.2. Finally, the rays that are propagating in the same direction are merged into one component whose intensity is defined by the sum of individual rays. For the reason of simplicity, we analyze the situation that corresponds to a selected ray, approaching a textured interface from the left side only (Figure 2.14). Each ray can be subsequently treated in the presented way by considering the change in the incident angle and the indexes "l" and "r."

The specular and diffused reflectance and transmittance of the interface, the angles of reflected and transmitted specular components of light, and the *ADFs* of the scattered parts have to be defined in order to calculate the intensities of reflected and transmitted light rays. Here, the diffused reflectance (defined for light intensity) can be determined straightforwardly by multiplying the total reflectance by the haze parameter as given in Equation (2.42)

$$R_{dif\,l}\left(\varphi_{inc}\right) = R_{0\,text\,l}(\varphi_{inc}) \cdot H_{R\,l}(\varphi_{inc}) \tag{2.42}$$

where $R_{0\,text\,l}(\varphi_{inc})$ can be equal to or lower than the reflectance $R_{0\,l}(\varphi_{inc})$ corresponding to the flat interface (due to antireflecting effect at a textured interface). As indicated in the equation, the haze parameter $H_{R\,l}(\varphi_{inc})$ is, in general, also dependent on the incident angle. Further on, specular reflectance $R_{spec\,l}(\varphi_{inc})$ can be determined by Equation (2.43)

$$R_{spec\,l}(\varphi_{inc}) = R_{0\,text\,l}(\varphi_{inc}) - R_{dif\,l}(\varphi_{inc}) = R_{0\,text\,l}(\varphi_{inc}) \cdot (1 - H_{R\,l}(\varphi_{inc})) \tag{2.43}$$

where it is considered that the sum of specular and diffused reflectances is equal to the total reflectance.

Similarly, the equations can be written for the diffused and specular transmittances

$$T_{\text{dif lr}}(\varphi_{\text{inc}}) = T_{0\text{ text lr}}(\varphi_{\text{inc}}) \cdot H_{T\text{ lr}}(\varphi_{\text{inc}}) \tag{2.44}$$

$$T_{\text{spec lr}}(\varphi_{\text{inc}}) = T_{0\text{ text lr}}(\varphi_{\text{inc}}) - T_{\text{dif lr}}(\varphi_{\text{inc}}) = T_{0\text{ text lr}}(\varphi_{\text{inc}}) \cdot (1 - H_{T\text{ lr}}(\varphi_{\text{inc}})) \tag{2.45}$$

where

$$T_{0\text{ text lr}}(\varphi_{\text{inc}}) = 1 - R_{0\text{ text l}}(\varphi_{\text{inc}}). \tag{2.46}$$

Based on defined $R_{\text{spec l}}(\varphi_{\text{inc}})$, $T_{\text{spec lr}}(\varphi_{\text{inc}})$, $R_{\text{dif l}}(\varphi_{\text{inc}})$, and $T_{\text{dif lr}}(\varphi_{\text{inc}})$, the intensities $I^-_{\text{R spec l}}$, $I^{--}_{\text{R scatt common l}}$, $I^+_{\text{T spec r}}$, and $I^+_{\text{T lr scatt common r}}$ can be determined as in Equations (2.47) through (2.50).

$$I^-_{\text{R spec l}} = R_{\text{spec l}}(\varphi_{\text{inc}}) \cdot I^+_{\text{inc l}} \tag{2.47}$$

$$I^-_{\text{R scatt common l}} = R_{\text{dif l}}(\varphi_{\text{inc}}) \cdot I^+_{\text{inc l}} \tag{2.48}$$

$$I^+_{\text{T spec r}} = T_{\text{spec lr}}(\varphi_{\text{inc}}) \cdot I^+_{\text{inc l}} \tag{2.49}$$

$$I^+_{\text{T scatt common r}} = T_{\text{dif lr}}(\varphi_{\text{inc}}) \cdot I^+_{\text{inc l}} \tag{2.50}$$

For specular components the directions (angles $\varphi_{R\text{ l}}$, $\varphi_{T\text{ lr}}$) are defined in the same way as in the case of a flat interface (following Snell's law; see Section 2.4.2). The angular distribution of the intensities of the scattered components is described by $ADF_{R\text{l}}$ and $ADF_{T\text{lr}}$, which are input parameters. In this case, the $ADF$s are functions not only of a scattering angle but also of the incident angle of the incoming light ray (see Section 3.2.6).

## 2.6  OPTICAL SITUATION AND CALCULATION PROCEDURE IN A MULTILAYER STRUCTURE

In this section we present one possible approach to "tracing" the specular and scattered light components in a multilayer structure with textured and flat interfaces. The propagation of light throughout the layers and the situations at the interfaces that were described in the previous sections must be combined in this procedure. Different approaches have been reported to tackle this problem. Matrix formalism can be used (Zeman et al. 1997), or individual waves and rays can be traced iteratively (Springer, Poruba, and Vaneček 2004). In the first

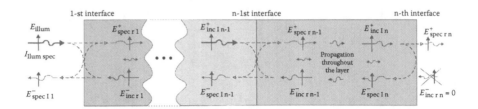

**FIGURE 2.15** Calculation procedure for coherent specular light in a multilayer structure with flat and textured interfaces.

case, the problem is transferred to solve matrix equations in order to determine the common light intensities of forward- and backward-going light waves and rays. The second option gives more insight into the propagation of the individual components inside the structure, but it can be time consuming if not tackled properly.

In the following, the approach that was also taken in the Sun*Shine* optical simulator is described in more detail. The incident illumination is assumed to be applied from the front (left) side of the structure only. Specular or diffused illumination can be used, where in the case of the diffused illumination its *ADF* also can be chosen, while specular light enters the structure perpendicularly to the interfaces. In the presented approach, the propagation of light is divided into three phases (Figures 2.15 and 2.16). In the first phase, the propagation of specular components throughout the structure is determined completely. In addition, scattered components at the textured interfaces are determined and partially also their propagation is considered within the first phase. The second and the third phase of calculation correspond to further propagation of the scattered light (ray tracing). These two phases are repeated iteratively as explained later.

In the first phase (Figure 2.15) the calculation starts at the right side of the last interface of the structure. The approach is based on Saeng-udom (1991).

First, an arbitrary non-zero value is applied to the electric field strength of the forward-going specular wave at the last, $n$-th interface ($E^+_{\text{spec r n}}$). Later on, this value will be scaled according to the actual intensity of the specular incident illumination. At the right side of the last interface, no backward-going wave is assumed (no backside illumination, $E^-_{\text{inc r n}} = 0$). By defining $E^+_{\text{spec r n}}$ and $E^-_{\text{inc r n}}$ the electric field strengths of the specular components at the left side of the interface, $E^+_{\text{inc l n}}$ and $E^-_{\text{spec l n}}$, can be calculated by applying the equations for a flat (Section 2.4.1) or a textured interface (Section 2.5.1), depending on the type of interface. If the interface is textured, the scattered light components at the left and right sides are determined (common intensities and *ADF*s).

In the next step of the first phase the propagation of $E^+_{\text{inc l n}}$ and $E^-_{\text{spec l n}}$ toward the preceding, ($n$-1)th interface, is applied, considering the corresponding equations from Section 2.3. It has to be pointed out that for the forward-going wave $E^+_{\text{inc l n}}$, the propagation is determined for its opposite direction; thus, in the case of an absorbing medium, the absolute value of $E^+_{\text{inc l n}}$ is increasing toward

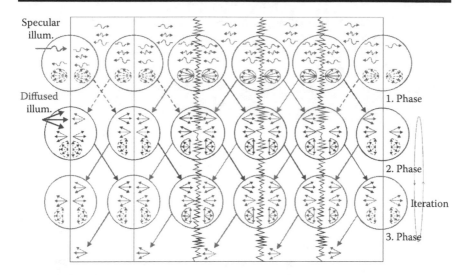

**FIGURE 2.16** Calculation procedure for coherent specular light and incoherent scattered light in a multilayer structure with flat and textured interfaces. The procedure is divided into three phases; the last two phases are repeated iteratively until the minimal trash-hold of intensity of scattered light has been achieved.

the $(n-1)$th interface. If there was the scattered light component determined at the right side of the $n$-th interface, the component is propagated toward the preceding interface as well in this phase (see Figure 2.16 where the first phase is represented by the first row of circles), according to the equations described in Section 2.3.3.

At the $(n-1)$st interface from $E^+_{spec\,r\,n-1}$ and $E^-_{inc\,r\,n-1}$ (actually $E^+_{inc\,l\,n}$ and $E^-_{spec\,l\,n}$ after propagation throughout the layer), the $E^+_{inc\,l\,n-1}$ and $E^-_{spec\,r\,n-1}$ and new scattered light components in the case of a textured interface are determined, considering the corresponding interface conditions. Then, the waves and scattered components are propagated toward the $(n-2)$th internal interface. The described procedure of calculation is repeated until the front (first) interface is achieved. The calculated $E^+$ and $E^-$ values at the interfaces (e.g., in the form of $E^+_{inc\,l\,n-1}$, $E^-_{spec\,l\,n-1}$, etc.) or inside a layer (at an arbitrary point $x$) already represent the sum of electric field strengths of all forward- and backward-going waves, respectively, that originate from multiple reflectances and transmittances at internal interfaces.

The calculated component $E^+_{inc\,l\,1}$ at the left side of the front interface has to be scaled in order to become equal to the electric field strength of the actual incoming specular light illumination, $E_{illum}$. The $E_{illum}$ can be determined from the known intensity of the incident light $I_{illum\,spec}$ at a chosen discrete wavelength as given in Equation (2.51).

$$E_{illum} = \sqrt{\frac{2I_{illum\,spec}}{Y_0 n}}$$
(2.51)

The scaling factor, *SF*, that has to be applied to the calculated $E^+_{inc\,11}$ at the left side of the front interface, and consequently to all electric field strengths determined in the structure, can be calculated as given in Equation (2.52).

$$SF = \frac{E_{illum}}{E^+_{inc\,11}} \tag{2.52}$$

Furthermore, the calculated intensities of scattered components that were determined and propagated throughout the layers thus far have to be scaled with the square of *SF*, considering the relationship between electric field and light intensity. With this step the first phase of the calculation procedure is completed.

The described procedure is changed slightly if, at a certain interface, the level of scattered light becomes unity and, consequently, the specular component becomes zero from this interface on. If such an interface is detected during the calculation of a specular component, starting at the rear interface toward the front interface, all the electric field strengths from this interface, where the specular component disappears, toward the rear interface have to be set to zero and the calculation starts again at this interface (in this case at its left side). In Figure 2.16, the procedure of calculation in the first phase is summarized (first row of circles including backward-pointing arrows above the second row). The circles indicate situations at the interfaces, whereas the propagations are denoted with wavy (coherent specular light) and straight arrows (incoherent scattered light).

In the second phase, the scattered light is traced in the forward direction only as shown in Figure 2.16 (second row of circles, including forward-pointing arrows above and below the second row). The diffused illumination component applied at the front interface (optional) is taken into the calculation procedure in the first iteration of this phase. Besides propagation, the optical situations at flat and textured interfaces also are determined (circles). Each ray (propagated direction) is analyzed individually. However, in applying the presented method in a numerical simulation, the directions in which the rays are propagating are discretized; thus, the rays that are propagating in the same direction cone defined by $\varphi_i +/-\Delta\varphi_{step}/2$ ($\varphi_i$ – *i*-th discrete angle of propagation, $\Delta\varphi_{step}$ – discrete angular step) can be combined in one common ray. The intensity of the common ray is determined by the sum of the intensities of the individual rays propagating in the same discrete cone. In this way the number of discrete rays is not increasing in the simulation.

In the second phase the input intensities of light for the third phase are determined. In the third phase (third row of circles in Figure 2.16, including backward-pointing arrows above and below the row) the rays are traced in a backward direction only. When the third phase is finished the inputs for the next iteration of the second phase are determined. The second and the third phase are repeated until all the intensities at the interfaces do not decrease (due to absorption or escaping out of the structure) down to a predefined minimal trash-hold level.

After the last iteration the common light intensity profile is defined throughout the structure by summing up the intensities of all forward-going rays (specular and diffused) at each discrete point in the structure (irrespective of their angle) and subtracting the sum of all backward-going rays at the same point. This common light intensity profile presents the basis for determination of the charge carrier generation rate profile $G_L(x, \lambda)$ (see Section 3.3).

## 2.7    IMPLEMENTATION OF THE MODEL IN OPTICAL SIMULATOR SUN*SHINE*

Based on the theoretical background described in the previous sections of this chapter, a 1-D semi-coherent optical simulator Sun*Shine* was developed (Krč, Smole, and Topič 2002, 2003). The details of Sun*Shine* can be found in the user's manual (Krč and Topič 2011); here, only some main features are highlighted. The simulator was developed for detailed analysis and investigation of optical effects in thin-film optoelectronic structures with flat and nano-textured interfaces. It was used mainly for thin-film photovoltaic devices (as presented in Chapter 5) but also for photodetectors and other devices (Krč 2002).

The Sun*Shine* is a useful tool in design, optimization, and development of novel optical concepts for thin-film solar cells (López, Vega, and López 2012). It also facilitates the study of optical properties of conventional, hetero-junction silicon or other types of solar cells where multiple thick layers can be used in the structure. However, at present the larger geometrical features of texturization, which would require the use of geometrical optics, cannot be analyzed directly with the simulator Sun*Shine*. Also, the simulator cannot be used directly to study in more detail the relationship between interface texture (morphological parameters) and light scattering at an interface. However, this link is established by using input scattering parameters ($H$ and $ADF$) that can be experimentally determined or theoretically calculated ahead of the simulations.

A graphical user interface of the simulator Sun*Shine* enables us to enter the input parameters of simulations (e.g., structure, optical properties of layers, scattering parameters, etc.) and to present the results of simulations in an easy, user-friendly way. Figures 2.17a and b show the main interface console where the structure can be defined and the results console, respectively.

To conclude, this chapter presented some background of optical situations, which can be considered and implemented in a 1-D optical model for simulation of thin-film photovoltaic devices. The presented knowledge about the optical situation at interfaces and inside layers can also help us to understand the situation if more comprehensive 2-D or 3-D modeling is used, where the physics is represented on the level of Maxwell equations. One of the challenges of the presented 1-D modeling is proper determination of input parameters—in particular, the descriptive scattering parameters for internal interfaces, which is the topic of the next chapter.

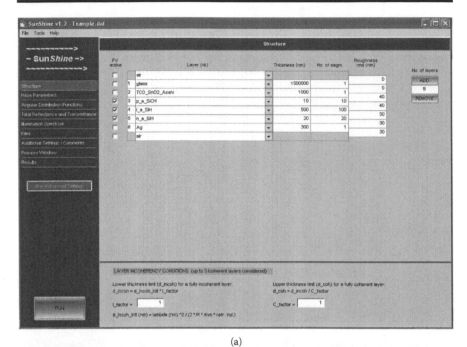

(a)

(b)

**FIGURE 2.17**    Graphical user interface of the Sun*Shine* simulator: (a) the structure console and (b) the results console.

# REFERENCES

Jung, S., K.-Y. Kim, Y.-I. Lee, J.-H. Youn, H.-T. Moon, J. Jang, and J. Kim. 2011. "Optical Modeling and Analysis of Organic Solar Cells with Coherent Multilayers and Incoherent Glass Substrate Using Generalized Transfer Matrix Method." *Japanese Journal of Applied Physics* 50 (12). doi:10.1143/JJAP.50.122301.

Kong, J. A. 1990. *Electromagnetic Wave Theory.* 2nd ed. Wiley-Interscience.

Krč, J. 2002. *Analysis and Modelling of Thin-Film Optoelectronic Structures Based on Amorphous Silicon.* PhD thesis, University of Ljubljana.

Krč, J., B. Lipovšek, and M. Topič. 2012. Light-Management in Thin-Film Silicon Solar Cell. In: Lopez, A. B. C., A. M. Vega, and A. L. Lopez, eds., *Next Generation of Photovoltaics: New Concepts*, pp. 95–130. Berlin: Springer.

Krč, J., F. Smole, and M. Topič. 2002. "One-Dimensional Semi-coherent Optical Model for Thin-Film Solar Cells with Rough Interfaces." *Informacije MIDEM—Journal of Microelectronics, Electronic Components and Materials* 32 (1): 6–13.

Krč, J., F. Smole, and M. Topič. 2003. "Analysis of Light Scattering in Amorphous Si:H Solar Cells by a One-Dimensional Semi-coherent Optical Model." *Progress in Photovoltaics: Research and Applications* 11 (1): 15–26. doi:10.1002/pip.460.

Krč, J., and M. Topič. 2011. "*SunShine* Optical Simulator V1.2.8—User's Manual." University of Ljubljana.

Ley, L. 1984. "Photoemission and Optical Properties." In *The Physics of Hydrogenated Amorphous Silicon II*, ed. J. D. Joannopoulos and G. Lucovsky, 61–168. Springer. http://www.springerlink.com/content/p153924824732407/.

Saeng-udom, R. 1991. *Optical Design of Single- and Multi-junction Solar Cells.* PhD thesis, University of Neubiberg.

Santbergen, R. 2008. *Optical Absorption Factor of Solar Cells for PVT Systems.* Eindhoven University Press.

Schropp, R. E. I., and M. Zeman. 1998. *Amorphous and Microcrystalline Silicon Solar Cells: Modeling, Materials and Device Technology.* Springer.

Springer, J., A. Poruba, and M. Vaneček. 2004. "Improved Three-Dimensional Optical Model for Thin-Film Silicon Solar Cells." *Journal of Applied Physics* 96 (9) (November 1): 5329–5337. doi:10.1063/1.1784555.

Stiebig, H. 1997. *Entwicklung und Beschreibung von optoelektronischen Bauelementen auf der Basis amorphen Siliziums.* PhD thesis, Berichte des Forschungszentrum Jülich.

Stiebig, H., A. Kreisel, K. Winz, M. Meer, N. Schultz, C. Beneking, T. Eickhoff, H. Wagner. 1994. "Spectral Response Modelling of a-Si:H Solar Cells Using Accurate Light Absorption ProfilEs." Proc. First World Conference on Photovoltaic Energy Conversion (WCPEC). Hawaii, Dec 5–9, 1994, pp. 603–606.

Troparevsky, M. C., A. S. Sabau, A. R. Lupini, and Z. Zhang. 2010. "Transfer-Matrix Formalism for the Calculation of Optical Response in Multilayer Systems: From Coherent to Incoherent Interference RID B-9571-2008." *Optics Express* 18 (24) (November 22): 24715–24721. doi:10.1364/OE.18.024715.

Zeman, M., J. A. Willemen, L. L. A. Vosteen, G. Tao, and J. W. Metselaar. 1997. "Computer Modelling of Current Matching in a-Si:H/a-Si:H Tandem Solar Cells on Textured TCO Substrates." *Solar Energy Materials and Solar Cells* 46 (2) (May): 81–99. doi:10.1016/S0927-0248(96)00094-3.

# Input and Output Parameters

**3**

## 3.1   INTRODUCTION

Input parameters of optical modeling and simulations are the data that determine the simulated optical structure, consisting of layers and interfaces in the case of thin-film solar cells, the surrounding medium (usually air), properties of the illumination source, and the settings of the calculation process. Output parameters are selected results obtained by simulation.

Determination and selection of the values of input parameters present a crucial task even for users who are specialists in the field. To carry out realistic simulations, it is essential that we use proper physical values of all input parameters in simulations. Theoretical models and experimental techniques are used to determine the input parameters. Some of them we highlight in this chapter.

However, it often happens that not all of the input parameters can be assigned to predefined realistic values. In these cases it is important that we estimate at least the intervals in which the values are expected. Using test structures, applying reverse engineering with fitting procedures can be used within the simulation process to estimate the unknown value of the free parameter (Schropp and Zeman 1998). Credibility of such fitting procedures and the values obtained must be established.

With respect to the input parameters, the freedom in (unlimited) variation of the input parameter gives us the opportunity to investigate possible device solutions beyond the current state-of-the-art devices and, in this way, gives directions for further improvements and new concepts of the devices.

After input parameters are determined and simulations have been carried out, we come to the output parameters and characteristics as results of simulations. They give us information about the analyzed device operation: not only external but also internal effects and properties are revealed. Despite proven reliability of models used and careful selection of input parameters, we must ascertain whether the obtained results are reasonable; that is, whether they really correspond to our understanding of device operation and expectations. Usually our simulated structures and devices are rather complex and their internal and external characteristics cannot be predicted fully; however, general correctness of the results has to be judged. If results cannot be explained, cross-checks on simplified structures, whose output characteristics can be calculated or estimated using some analytical models, is appreciable, especially when simulating new concepts of devices with less predictable behavior.

This chapter introduces the reader to the main input and output parameters of optical simulations. The emphasis is on the one-dimensional (1-D) semi-coherent optical approach described in Chapter 2, which requires more input parameters than, for example, two-dimensional (2-D) and three-dimensional (3-D) rigorous modeling approaches, in which light scattering can be determined directly from interface morphologies. Selected experimental and analytical methods of determination of input parameters are described in the chapter. Determination of light-scattering parameters of nano-textured interfaces that present one of the most challenging issues in 1-D modeling and simulation of thin-film solar cells is presented in more detail.

After sections devoted to input parameters, the chapter focuses on main output parameters of optical simulations. In the very last part of the chapter, we present some selected examples and results of calibration of the 1-D semi-coherent optical simulator Sun*Shine* with input parameters corresponding to specific textured substrates and different types of thin-film solar cells.

## 3.2  INPUT PARAMETERS

Main input parameters of optical simulations of thin-film solar cells define

- the illumination spectrum
- the multilayer structure with the definition of individual layer thicknesses and their optical properties
- properties of interfaces, especially textured ones, including morphological or light-scattering characteristics and other optical properties

In the following, we explain each of these points in more detail. We describe how to determine the input parameters using both experimental and theoretical methods.

### 3.2.1  ILLUMINATION PROPERTIES

Basic properties of illumination, which are important for solar cell simulations, are

- illumination power density as a function of light wavelength (illumination spectrum)
- fraction of direct and diffused light in the spectrum
- the incident angle of direct illumination
- the angular distribution of the diffused light

For the direct light we assume here that it is a plane wave with negligible divergence (our illumination source, the sun, is far away from the solar cell). On the contrary, the diffused light is reflected and scattered from the clouds and objects on the earth and has multiple directions.

Typically, photovoltaic (PV) devices, especially laboratory solar cells, are first tested indoors under standard test conditions (IEC 60904-3:2008). A solar simulator is used as an artificial light source, resembling the spectrum of the sun, which is characterized by an AM 1.5 solar spectrum. Mostly a homogenous and collimated direct illumination is present at indoor testing. When we consider outdoor conditions, the illumination properties can be different in several aspects. To include these variations in optical simulations as well, all of the previously stated input parameters need to be included in the simulation.

The first parameter—the power density of the light source—is in numerical simulations discretized into wavelength intervals (step of 10 nm or less). For each interval we calculate the corresponding power density in W/m$^2$ or mW/cm² by integrating the spectrum density [e.g., in mW/(cm²µm)] over the chosen discrete wavelength interval $\Delta\lambda$.

$$I(\lambda_i) = \int_{\lambda_i - \frac{\Delta\lambda}{2}}^{\lambda_i + \frac{\Delta\lambda}{2}} p(\lambda)\cdot d\lambda \qquad (3.1)$$

An example of such a discretized spectrum is shown in Figure 3.1 for the case of the AM 1.5 spectrum, considering the interval $\Delta\lambda = 10$ nm. These data present direct input for optical simulations.

The possibility to include different illumination spectra in the simulations enables us to investigate the behavior of the PV devices under realistic outdoor conditions or under different artificial illumination sources. The effects of different spectra are of special interest for multi-junction devices, where the current matching between cell components can be affected.

As mentioned, in addition to direct illumination the diffused component is present under realistic outdoor conditions. To describe the fraction of direct and diffused light we can use a diffusion factor (the ratio between diffused and

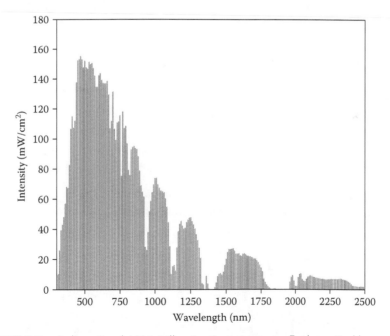

**FIGURE 3.1**  A discretized AM 1.5 illumination spectrum. Each vertical bar corresponds to the intensity of 10 nm broad wavelength interval in the spectrum.

total [direct + diffused] component). Furthermore, the direct component in a realistic case can approach the surface of the PV device under a different incident angle, depending on the position of the sun on the sky and PV device orientation. Different angular distributions are expected also for the diffused component of the illumination spectrum. We need to consider these angular dependencies in optical simulations if we want to study the behavior of the devices under outdoor conditions.

In the simulator Sun*Shine* (see Chapter 2), for example, the incident illumination can have both direct and diffused components. The direct component can be treated as coherent light, but only if it is applied under a perpendicular incident angle. Illumination under other incident angles, being either direct or diffused, can be treated as incoherent light only in this case. This is one of the drawbacks related to most semi-coherent optical models. Regarding the diffused light in the Sun*Shine*, the user can use the illumination with adjustable diffusion factors (0%–100%), considering different user-defined angular distribution functions (*ADFs*) of the diffused component. The diffused light is analyzed incoherently.

In case of rigorous 2-D and 3-D optical simulations, which is the topic of Chapter 4, all light is treated as coherent light (electromagnetic waves). When propagating throughout thick layers, like glass at the front, dense interference may arise in output characteristics, as mentioned in Section 2.3.2. In these simulations it is still a challenge to describe incoherent properties of light in an effective manner.

### 3.2.2   STRUCTURE OF THE DEVICE

For the general description of the device structure, the following basic parameters have to be defined:

> the type and the sequence of layers (starting with the incident medium, following with layers of the structure, and ending with the outgoing medium),
> dimensions and the discretization of layers (the latter is always needed in numerical simulation),
> the side of applied incident illumination, and
> the type of interfaces (flat or textured).

The sequence of layers is important from the light propagation point of view. It also defines the layer pairs that form the interfaces in the structure. We discretize each layer in the $x$-axis (vertical) direction in the case of 1-D simulations and in the other two (lateral) directions if we carry out 2-D or 3-D simulations. In this way the quantities such as carrier generation rate profile or light intensity profile across the layer can be determined. At this point the interfaces can be divided in two groups: flat and textured. In later steps we define optical properties of the interfaces, including scattering for textured interfaces.

In Figure 3.2a we present a general scheme of a thin-film multilayer structure and in Figure 3.2b a schematic view of a realistic amorphous silicon single-junction solar cell in a superstrate configuration, as an example of multilayer structure of our interest. The superstrate configuration means that the substrate of the structure is located at the front side of the structure with respect to the light incidence. Therefore, it is called superstrate. As the superstrate, typically a glass plate is used, ensuring high transmission properties that are required to enable the light to enter the basic solar cell structure. From the optical point of view the structure shown in Figure 3.2b is composed of a thick glass superstrate (thickness in the range of millimeters), followed by a five-layer thin-film structure (thickness of layers in the range of nanometers and micrometers). The first layer in the solar cell structure that follows the glass is the transparent conductive oxide (TCO) layer, which represents the front electrical contact of the cell. The TCO layer is also used to introduce textured interfaces to internal interfaces of the solar cell structure, as indicated in Figure 3.2b. Through the TCO superstrate, light enters the basic solar cell structure, consisting of $p$-doped, intrinsic ($i$-), and $n$-doped a-Si:H layers in the case of amorphous silicon solar cell structure. The $i$-layer presents the absorber layer in which most of the light is aimed to be absorbed. After the $n$-layer a silver layer, representing the back contact, is added to complete the solar cell structure. Often, a rear TCO (undoped zinc oxide) is added to improve the reflectivity properties of the back contact.

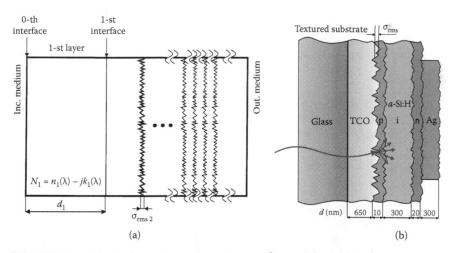

**FIGURE 3.2** (a) A schematic representation of a multilayer thin-film structure. (b) A single-junction amorphous silicon solar cell as an example of a multilayer structure.

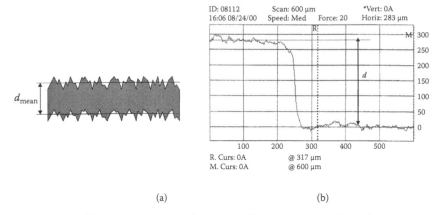

(a)                                    (b)

**FIGURE 3.3**   (a) Mean thickness of a textured layer that is used in 1-D simulations. (b) An example of a profilometric measurement of layer thickness.

### 3.2.3   LAYER THICKNESS

Usually, in simulated structures of thin-film solar cells we assume a high level of homogeneity of layer thicknesses in the device. Although in the real world certain deviations exist, especially if we consider large-area modules, in simulations often an average single thickness value is assigned to each layer. In the case of layers with textured interfaces, we can use the distance between the mean levels of the top and the bottom interface as a good estimate of the layer thickness in simulations (see Figure 3.3a).

To determine the thickness of thin layers used in thin-film solar cells, we can apply different characterization techniques. First, the thickness can be estimated from the growth rate and the time of deposition when processing the layer. Next, we can apply contact or non-contact profilometric measurements across the cut edge of a layer to determine the thickness (Rodriguez, Austin, and Bartlett 2012). An example of contact profilometric measurement is shown in Figure 3.3b. Optical methods like ellipsometry and reflectance-transmittance measurements can also be applied to determine the thickness; however, in these cases we have to know the optical properties of the films (Tompkins 2006).

In thin-film PV devices we combine layers of relatively small (range of nanometers) and relatively large thicknesses (superstrates and substrates are in the range of millimeters). From the optical point of view, this requires different optical treatments of the layers in simulations (coherent/incoherent) as discussed in Section 2.3.

### 3.2.4   OPTICAL PROPERTIES OF LAYERS

In simulations, optical properties of layers are often described by complex refractive indexes $N(\lambda) = n(\lambda) - jk(\lambda)$, or by complex dielectric (permittivity) functions $\varepsilon(\lambda) = \varepsilon_1(\lambda) - j\varepsilon_2(\lambda)$ as mentioned in Section 1.3 of Chapter 1.

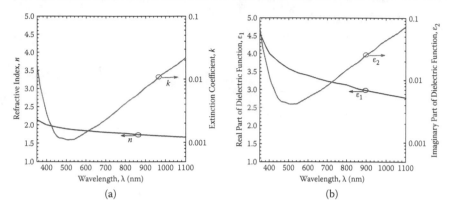

**FIGURE 3.4**    (a) Refractive index and extinction coefficient of a ZnO:Al TCO layer (not optimized) and (b) the corresponding relative dielectric functions $\varepsilon_1$ and $\varepsilon_2$ as a function of the wavelength.

Wavelength dependency of optical properties of materials has to be considered if a broad spectrum of light, as in the case of solar cells, is taken into account. Examples of wavelength dependent $n(\lambda)$, $k(\lambda)$, and of the corresponding $\varepsilon_1(\lambda)$ and $\varepsilon_2(\lambda)$ are shown in Figure 3.4 for an aluminum-doped magnetron sputtered zinc oxide material (ZnO:Al), which we can use as a front TCO electrode in thin-film silicon solar cells. The analytical relation between the complex refractive index and permittivity function was given in Equation (1.4). For the ZnO:Al example shown, optical properties depend significantly on the Al doping level and on conditions of sputtering. With lower doping, one can decrease the long-wavelength absorption ($k(\lambda)$) by decreasing the concentration of free carriers (Agashe et al. 2003). On the other hand, the electrical conductivity of TCO materials is of prime importance, which is increased by higher doping.

As for the thicknesses we often assume also for optical properties of layers a high level of vertical and lateral homogeneity in a device in simulations. For the realistic layers, which do not follow this assumption, we can use average values of the deviations in lateral directions, whereas for vertical deviations we can define sub-layers with different optical properties. Most often this does not need to be considered for state-of-the-art small-area (e.g., $1 \times 1$ cm$^2$) devices.

Accurate determination of optical properties of realistic thin-film materials used in PV technology is of great importance for reliable optical simulations, since material properties depend on deposition conditions (controlled and uncontrolled). General optical data obtained from the literature are often not accurate enough to carry out optical simulations of our devices. It is important that we determine optical properties of the actual layers used in the device.

Accurate determination of the refractive index, and especially the extinction coefficient in the wavelength regions where its values are low, presents a challenge. Different methods of determination of optical properties of thin films can be used and combined. Most commonly are ellipsometry (Tompkins 2006), reflectance-transmittance measurements (R-T) (Bie et al. 2000; Peiponen, Myllylä, and Priezzhev 2010; Sap et al. 2011), constant photocurrent method (CPM) (Schmidt et al. 2000; Holovský et al. 2008), photo-deflection spectroscopy (PDS) (Gracin et al. 2009), and Fourier transform infrared (FTIR) spectroscopy (Vaneček and Poruba 2002). Details of these methods can be found in the references listed at the end of the chapter.

As an example of a method for determination of optical properties (and layer thicknesses) of thin semi-transparent films we briefly present an approach based on R-T measurements. In the presented approach refractive index and extinction coefficient as functions of light wavelength can be determined even if the dispersion models for complex dielectric function $\varepsilon$ of the specific material are not established. The method is often sufficiently accurate to determine optical properties of thin layers for the purpose of their further use in optical simulations on the device level. For more accurate determination of optical properties, which is sometimes required when studying material properties (e.g., defect states) in the wavelength region of low absorption, we should use more reliable methods.

In the presented R-T approach, $N(\lambda)$ is determined directly from the reflectance, $R(\lambda)$, and transmittance, $T(\lambda)$, measurements of the characterized film on a transparent substrate. The key issue presents the extraction of the realistic solutions of $N(\lambda)$ from the pool of mathematically possible solutions, as explained later. First, we present an analytical optical model of a typical sample structure (Figure 3.5a) used for the determination of optical properties of

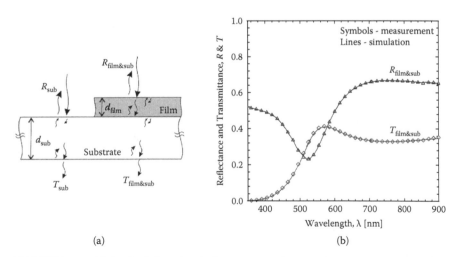

(a)                                                    (b)

**FIGURE 3.5**   (a) Optical situation (reflectance and transmittance components) at the substrate and thin-film/substrate optical system. (b) Measured and simulated reflectance and transmittance of the a-Si:H film (53.7 nm)/glass (1 mm) optical system. Good agreement between measured and simulated data confirms the validity of the determined complex refractive indexes of the substrate and the film.

a film. The model includes a transparent substrate (usually glass) and a semi-transparent thin film. Since $N_{sub}$ ($\lambda$) of the actual glass substrate used is often also not accurately known (there exist several types of glass substrates), we have to define it first. Thus, the model has to be applicable also to the glass substrate alone. Due to the large thickness of the glass, incoherent analysis of light in the substrate has to be considered in the model. For a thin layer under investigation we have to apply, on the other hand, a coherent analysis of light. Based on transfer matrix formalism (Troparevsky et al. 2010), we can derive the following analytical formulas to describe reflectance, $R_{film\&sub}$ and transmittance, $T_{film\&sub}$, of the entire film/substrate optical system [Equation (3.2)] (Macleod 2010). The quantities $R$ and $T$ correspond to reflectances and transmittances defined on the level of light power densities (intensities).

$$R_{film\&sub} = R_f + \frac{T_f T_f^- R_s e^{-8\pi \frac{k_{sub} d_{sub}}{\lambda}}}{1 - R_f^- R_s e^{-8\pi \frac{k_{sub} d_{sub}}{\lambda}}} \qquad T_{film\&sub} = \frac{T_f T_s e^{-4\pi \frac{k_{sub} d_{sub}}{\lambda}}}{1 - R_s R_f^- e^{-8\pi \frac{k_{sub} d_{sub}}{\lambda}}} \qquad (3.2)$$

$R_s$ and $T_s$ are reflectance and transmittance of the substrate/air interface and are defined by $N$ of free space, $N_{air} = 1.0 - j0.0$, and the $N$ of the substrate, $N_{sub}$, as shown in Equation (3.3). The $\lambda$ corresponds to the wavelength in free space.

$$R_s = \left| \frac{N_{air} - N_{sub}}{N_{air} + N_{sub}} \right|^2 \qquad T_s = 4 \frac{\text{Re}[N_{air} N_{sub}]}{|N_{air} + N_{sub}|^2} \qquad (3.3)$$

The minus superscript in some of the reflectance and transmittance denotations in Equation (3.2) indicates the reflectance and transmittance determined at the opposite side of the interface or film. For example, $R_s$ corresponds to the reflectance of the air/substrate interface (light applied from air), whereas the $R_s^-$ is the reflectance of the substrate/air interface (the same interface, but light applied from the substrate). However, for a single interface one can derive from Equation (3.2) that $R_s = R_s^-$ and $T_s = T_s^-$. This equality does not hold for $R_f$ and $T_f$ since these two quantities present effective reflectance and transmittance of the film as defined in Equation (3.4), where the matrix with cosine and sine elements is assigned to the characteristic matrix of the film (Macleod 2010).

$$R_f = \left| \frac{N_{air}B - C}{N_{air}B + C} \right|; \quad T_f = 4 \frac{\text{Re}[N_{air} N_{sub}]}{|N_{air}B + C|^2}; \quad \begin{bmatrix} B \\ C \end{bmatrix} = \begin{bmatrix} \cos\delta & j\frac{\sin\delta}{N_{film}} \\ j\sin\delta N_{film} & \cos\delta \end{bmatrix} \begin{bmatrix} 1 \\ N_{sub} \end{bmatrix}$$

$$(3.4)$$

The symbol delta is defined as $\Delta = 2\pi N_{\text{film}} d_{\text{film}}/\lambda$. To determine $R_f^-$ and $T_f^-$, $N_{\text{air}}$ and $N_{\text{sub}}$ have to be exchanged in Equation (3.4). By following Equations (3.2) through (3.4) we arrive at the system of non-linear equations symbolically represented by Equation (3.5),

$$R_{\text{film\&sub}} = R_{\text{film\&sub}}(N_{\text{sub}}, N_{\text{film}}) \quad T_{\text{film\&sub}} = T_{\text{film\&sub}}(N_{\text{sub}}, N_{\text{film}}) \qquad (3.5)$$

where $R_{\text{film\&sub}}$ and $T_{\text{film\&sub}}$ are measured quantities and $N_{\text{film}}$ (and $N_{\text{sub}}$) are unknown variables. To determine $N_{\text{sub}}$, first Equation (3.2), which in the given form corresponds to the entire film/substrate system, has to be applied to the substrate itself. This can be done by simple exchange of $R_{\text{film\&sub}}$ and $T_{\text{film\&sub}}$ with $R_{\text{sub}}$ and $T_{\text{sub}}$ (which are also measured quantities) and $R_f^-$ and $T_f^-$ with $R_s^-$ and $T_s^-$, respectively. The system of two non-linear equations including two unknown variables, which are $n$ and $k$—for the case of substrate and for the case of thin film after $n_{\text{sub}}$ and $k_{\text{sub}}$ are known—can be solved by using the Newton-Raphson algorithm (Kelley 1987). Multiple solutions are found (multiple pairs of $n$ and $k$ for the same wavelength) that all fulfill the equations for $R$ and $T$. By changing the starting approximation in the Newton-Raphson algorithm, we can arrive at these different solutions. Thus, special attention must be paid when using this method to extract the correct physical solution, represented by a single $n$ and $k$ pair at a single wavelength, of the material.

Wavelength-dependent $R_{\text{sub}}$, $R_{\text{film\&sub}}$, $T_{\text{sub}}$, and $T_{\text{film\&sub}}$ can be determined by means of a spectrophotometer. In our case, the PerkinElmer Lambda 950 spectrophotometer (PerkinElmer 2012) was used. To demonstrate the applicability of the method to the layers used in thin-film PV devices, here we present the results obtained for Corning glass substrate and i-a-Si:H layer deposited by plasma-enhanced chemical vapor deposit (PECVD) process. Samples with different thicknesses of films were fabricated at Tu Delft. We optically measured them in the wavelength range from 350 to 900 nm. Here, only the results obtained for a 53.7-nm thick a-Si:H film on 0.65-mm thick glass substrate (Corning 7059) are presented. $R_{\text{film\&sub}}$ and $T_{\text{film\&sub}}$ we show in Figure 3.5b. The accuracy of the thickness of the film is discussed later in this section.

The calculated multiple solutions (two in this case) for $n_{\text{sub}}$ are presented in Figure 3.6a and for extinction coefficient $k_{\text{sub}}$ in Figure 3.6b. First, we extract the realistic physical solution of $n_{\text{sub}}$. Here, the only possible realistic solution is the upper one in Figure 3.6a, where $n_{\text{sub}} > 1$, since $n$ of all solid materials should be greater than one (Palik 1991). After determining $n_{\text{sub}}$, and considering this value in the equations, we can extract the corresponding $k_{\text{sub}}$ (Figure 3.6b). The approximation curve between the point of $k_{\text{sub}}$, which was used in further analysis, is plotted in addition (dashed line).

All solutions of $n_{\text{film}}$ and $k_{\text{film}}$ of the characterized a-Si:H film we present in Figure 3.7a and 3.7b, respectively. The realistic solution for $n_{\text{film}}$ is, again, determined first. In contrast to the glass substrate case, the situation here is more complex, since there are several values of $n_{\text{film}}$ at single wavelengths, in general. Our procedure to extract the realistic solution starts at the longest

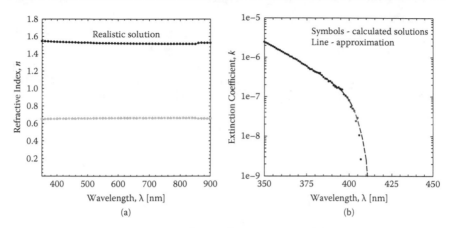

**FIGURE 3.6**    (a) The solutions of the refractive index of the analyzed glass substrate. (b) The solution of the extinction coefficient of the glass substrate obtained by considering the realistic solution for n. Note different scales for the wavelength in the figures.

wavelength (900 nm) and ends at the shortest wavelength (350 nm). Despite that the models of dielectric functions of the material are not needed here, still we have to estimate the region of expected values of refractive index for certain materials in order to determine the starting value for $n_{film}$ at longer wavelengths. For a-Si:H material the typical values of $n_{film}$ are between 3 and 5 in our wavelength region of interest (Springer, Poruba, and Vaneček 2004). In the 53.7-nm-thick a-Si:H film, however, there exists only one solution for $n_{film}$ at the longer wavelengths as can be observed when $\lambda \approx 900$ nm

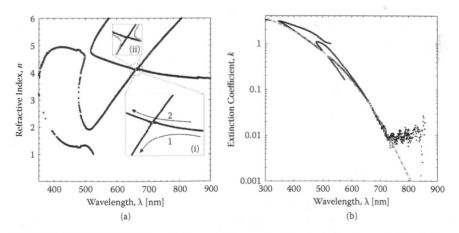

**FIGURE 3.7**    (a) Multiple solutions of the refractive index $n_{film}$ and (b) of the extinction coefficient $k_{film}$ calculated for 53.7-nm thick a-Si:H film on the Corning glass substrate.

in Figure 3.7a. This solution is automatically assigned to the physical solution, since its value is in the interval of expected solutions. By proceeding with calculation from longer to shorter wavelengths, several crossings of solutions may appear (two in the presented case). To determine the correct way to proceed at these crossings, the following aspect is considered in our procedure: for realistic materials the wavelength continuation (persistence) of $n$ is expected (Palik 1991). Thus, fast drops or increases are excluded from the analysis. According to the mentioned criteria, the inset (i) in Figure 3.7a shows the correct way (2) to proceed with the selection of correct $n_{film}$ at the first right crossing ($\lambda \approx 650$ nm). By following the wavelength continuation at the second crossing ($\lambda \approx 500$ nm), the criteria require a crossover to further solutions. Turning up in the curve, for example, would lead directly to two solutions at the same wavelength (which is not allowed), while turning down, on the other hand, does not satisfy the criteria for the wavelength continuation.

Additional analysis of the solutions in the region, where the crossing appears, showed that the upper and lower solution curves are shifted apart significantly, if the correct realistic thickness of the film, $d_{film}$, is not considered in the calculation as an input parameter. We found this observation to be a very useful and sensitive qualitative indicator for accurate determination (tuning) of the film thickness (up to 0.1 nm in the analyzed case). However, the accuracy of the R-T measurements is a crucial point here. In the inset (ii) in Figure 3.7a we show, for example, the curves at the right crossing if a thickness of 55 nm instead of 53.7 nm is used in the calculation (gray curves). A noticeable shift of the solution curves is observed in this case.

Once the physical solution for $n_{film}$ has been extracted, one can easily find the corresponding $k_{film}$ pairs from all solutions of $k_{film}$ presented in Figure 3.7b. Further analysis of all solutions of $k_{film}$ revealed that the solutions corresponding to $\lambda > 720$ nm contain noticeable numerical error due to very low absorption in the a-Si:H material in this wavelength region. Therefore, for this part extrapolation of correct solutions from shorter wavelengths is assumed as an acceptable approximation (dashed line). For a more accurate determination, other, previously mentioned methods have to be used. However, the obtained physical solutions for $n_{sub}$, $k_{sub}$, $n_{film}$, and $k_{film}$ were imported into the Sun*Shine* optical simulator to recalculate measured $R_{film\&sub}$ and $T_{film\&sub}$. The comparison is shown in Figure 3.6b. Almost perfect agreement between calculated (lines) and measured data (symbols) is observed in this case.

### 3.2.5   TEXTURED INTERFACE MORPHOLOGY

Textured interfaces are introduced in thin-film multilayer structures to increase the optical path in absorber layers by scattering the incident light at the interface corrugations. At the same time, textured interfaces can act as antireflecting structures at front interfaces of PV devices. The texturization on the surface of a layer can be achieved spontaneously in a grain-like layer growth (poly- and microcrystalline materials) or can be post-fabricated by different etching (Kluth et al. 2003) or embossing techniques (Escarré et al. 2011). In the first case the texture exhibits a random character of feature sizes and their distribution, whereas with the post-fabrication techniques either random or periodic character of the texture can be achieved.

For characterization of surface textures of interfaces/layers within thin-film solar cells, scanning electron microscopy (SEM) and atomic force microscopy (AFM) are commonly used. By means of SEM we can get a nice representation and visualization of realistic surface nano-textures (see examples in Section 3.5). In addition, with the cross-sectional images, the morphology of internal interfaces in a device can be detected. The AFM technique, on the other hand, enables us to quantitatively determine the morphology in terms of surface points $z(x, y)$ at different lateral locations (Figure 3.8).

In case of a random character of surface texture, statistical parameters are typically used to describe the morphology. Most common are

- the vertical root-mean-square roughness, $\sigma_{rms}$,
- the lateral correlation length, $L$,
- the root-mean-square slope of the surface profile, $s$,
- the height distribution, $h$,
- and others (average peak-to-peak height, average slope).

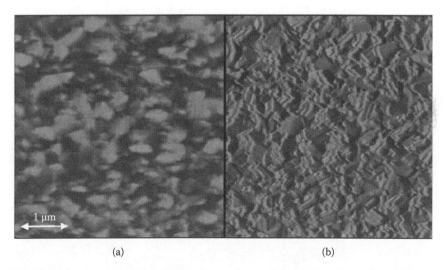

(a)                                        (b)

**FIGURE 3.8**  Example of an AFM image—the surface of a polycrystalline Cu(In,Ga) Se2 layer: (a) AFM height representation (lighter shades correspond to higher points). (b) AFM declination image of the same surface (first derivative of the surface function). It gives a better visualization of the vertical changes on the surface.

From a 1-D AFM profile $z(x)$ taken as a line scan at a textured surface, the $\sigma_{rms}$ parameter can be calculated as

$$\sigma_{rms} = \sqrt{\frac{1}{N}\sum_{i=1}^{N}\left[\overline{z} - z(x_i)\right]^2} \tag{3.6}$$

where $\overline{z}$ is the mean value (zero level) of the surface profile and $z(x_i)$ are the discrete points of the profile. Schematically the $\sigma_{rms}$ parameter is represented in Figure 3.9. It presents an effective value of the surface spatial signal. In the case of a 2-D surface scan, the sum of all points $z(x_i, y_j)$ in both lateral directions has to be considered.

In the optical simulator Sun*Shine*, for example, the surface morphology of a textured interface is represented only by $\sigma_{rms}$. Other surface parameters affecting scattering characteristics are involved indirectly through the descriptive scattering parameters, as explained later.

The correlation length $L$ is defined as a value where the autocorrelation function $G(l)$ [Equation (3.7)] of the profile is equal to k/e (e = 2.7183) (Bennett and Mattsson 1999).

$$G(k) = \frac{1}{N}\sum_{i=1}^{N-k} z(x_i) \cdot z(x_{i+k}) \tag{3.7}$$

where $k = 0 - N - 1$ and presents the length-shift between the two profiles $z(x)$. $L$ gives an indication of the average lateral feature size (period) of the texturization (see Figure 3.10).

The root-mean-square slope of the surface profile $s$ is defined as

$$s = \sqrt{\frac{1}{N}\sum_{i=1}^{N}\left[\frac{z(x_{i+1}) - z(x_i)}{\Delta x}\right]^2} \tag{3.8}$$

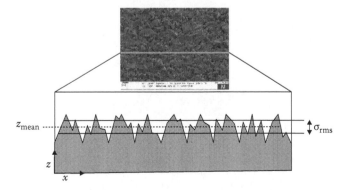

**FIGURE 3.9** Schematic representation of $\sigma_{rms}$ of a random nano-textured interface.

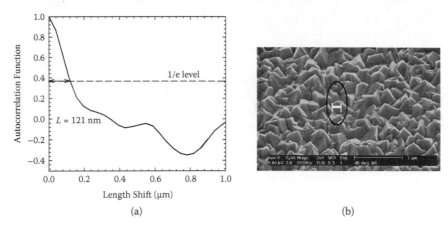

(a)                                                                  (b)

**FIGURE 3.10**    (a) An example of the autocorrelation function and denotation of the correlation length $L$ (SnO$_2$:F TCO surface). (b) The $L$ value is denoted also on the corresponding SEM image of the analyzed surface, indicating the average lateral feature size (period).

and gives information about the effective value of the steepness of the texture corrugations.

The height distribution $h$ gives quantitative information about the presence of different height levels in the surface profile (Figure 3.11). The height is measured from the mean level of the surface to the specific point in the profile.

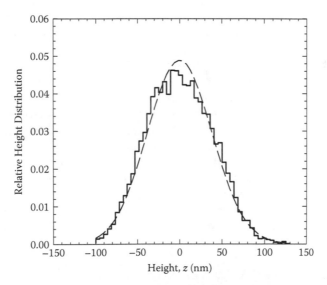

**FIGURE 3.11**    Example of a relative height distribution function (for SnO$_2$:F TCO surface shown in Figure 3.10). Gaussian function is plotted in addition (dashed curve), characterizing well the distribution of heights for the analyzed sample.

In some approaches a discrete Fourier analysis is used to investigate the textured surface morphology (Isabella, Krč, and Zeman 2010). Here, the surface profile is de-convoluted to basic sinusoidal components of different amplitudes, spatial frequencies, and phase shifts. The final surface can be represented as a sum of these components. The spatial frequency $f$ is defined as $f = 1/P$, where $P$ is the lateral period of a given sinusoidal component. This representation is essential to correlate surface morphology with light-scattering properties in some scattering models (Jäger and Zeman 2009; Dominé et al. 2010). Instead of showing the amplitude spectrum of the Fourier transform, the $\sigma_{rms}^*(f_i)$ spectrum can be determined (Isabella et al. 2010) by dividing the values of the components of the amplitude spectrum by $\sqrt{2}$ (which is the connection between the amplitude and the effective signal value, i.e., $\sigma_{rms}$ parameter for a sinus). We use an asterisk in the denotation $\sigma_{rms}^*(f_i)$ to differentiate between the function of *rms* roughness spectral representation and the $\sigma_{rms}$ parameter as one value. The $\sigma_{rms}$ parameter we can calculate from the $\sigma_{rms}^*(f)$ function as given in Equation (3.9) for a discretized case,

$$\sigma_{rms} = \sqrt{\sum_{i=1}^{M} \sigma_{rms}^*(f_i)^2} \tag{3.9}$$

where $M$ is the number of discrete spatial frequency components.

An example of $\sigma_{rms}^*(f)$ spectrum is presented in Figure 3.12 (Isabella et al. 2010).

Surface texture in combination with optical properties of the incident medium and the medium in transmission, together with the properties of incident light, determine the scattering and antireflective properties of light that is reflected from or transmitted through the interface. For antireflection effect, texture features in the sub-wavelength range (vertical and lateral dimensions smaller than light wavelength) or very large features leading to geometrical optics are most effective (Hagemann et al. 2008). For light scattering the (non-metallic) features have to be equal to or greater than the wavelength of light (Beckmann and Spizzichino 1987). Next we focus on light-scattering properties, which present a key issue for efficient light trapping in the solar cell structures. In the following sections scattering of light at randomly nano-textured interfaces, which are the most common in thin-film solar cells, will be investigated and discussed. In this case light scattering can be described by the descriptive scattering parameters such as haze parameters and *ADF*s, as mentioned in the two previous chapters.

### 3.2.6   LIGHT-SCATTERING PARAMETERS

We often use scattering parameters to characterize and describe light scattering at a nano-textured interface. By introducing scattering parameters into

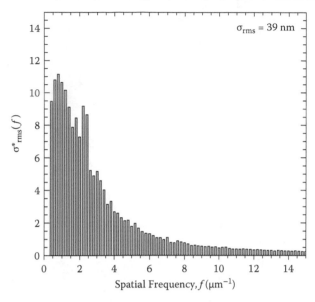

**FIGURE 3.12**  $\sigma^*_{rms}(f)$ spectrum of the SnO$_2$:F textured surface shown in Figure 3.10. The value of the $\sigma_{rms}$ parameter (calculated from $\sigma^*_{rms}(f)$ by using Eq. 3.9) is given on the top.

optical simulations we can make a separation of (1) the process of determination of scattering properties at the textured interfaces, which can be done outside of the multilayer system model, and (2) the simulation of light propagation in the multilayer structure, including textured interfaces. Accurate determination of light scattering at nano-textured interfaces usually involves comprehensive scattering theories in combination with different experimental methods. In the previously mentioned approach, once the scattering parameters are defined externally we can import and use them in the model of the multilayer system, avoiding the comprehensive scattering calculations inside this model. This is especially convenient for fast 1-D simulations, whereas in 2-D and 3-D rigorous approaches, scattering is implicitly determined inside simulations and no external parameters are needed. In the case of 1-D simulations theoretical approximations of scattering have to be used inside the propagation model to adapt the external scattering parameters exactly to the interfaces inside the multilayer structure. In these cases the imported scattering parameters, which are often measured in surrounding air, are often used to calibrate the scattering equations that are applied in multilayer system models to properly adapt the scattering properties to the morphology and optical surroundings of internal interfaces.

The purpose of defining the scattering parameters is also to simply evaluate and compare scattering abilities of surface-textured superstrates or substrates for photovoltaic applications. As we mentioned in previous chapters, there are two main types of scattering parameters: the haze parameters and the *ADF*s. In this chapter we present examples of experimental and theoretical approaches to the determination of scattering parameters. Application of scattering parameters to the interfaces inside thin-film solar cells (internal interfaces) in 1-D modeling is discussed.

First, we describe two commonly used measurement techniques for determining the haze parameter and *ADF*s:

- total integrating scattering measurement for determining the haze parameter and
- angular resolved scattering measurement for determining *ADF*s.

### 3.2.6.1   TOTAL INTEGRATING SCATTERING MEASUREMENTS

As mentioned in Section 2.5, we can define the haze parameter for reflected light, $H_R$, and transmitted light, $H_T$, of a textured surface as the ratio of diffused and total reflectance or transmittance:

$$H_R = \frac{R_{dif}}{R_{tot}} \qquad H_T = \frac{T_{dif}}{T_{tot}} \tag{3.10}$$

All of the quantities in Equation (3.10) are wavelength dependent in general.

To determine diffused and total reflectance ($R_{dif}$, $R_{tot}$) and transmittance ($T_{dif}$, $T_{tot}$) of a surface-textured substrate, we can use total integrating scattering (TIS) measurements. Basically, the TIS measuring system consists of a spectrometer equipped with an integrating sphere to collect the diffused light (Figure 3.13).

$T_{tot}$, $T_{dif}$, $R_{tot}$, and $R_{dif}$ are measuring using methods shown in Figure 3.14.

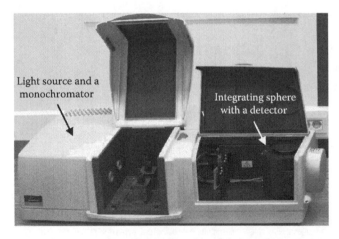

Light source and a monochromator

Integrating sphere with a detector

**FIGURE 3.13**   A spectrometer (PerkinElmer Lambda 950) with an integrating sphere.

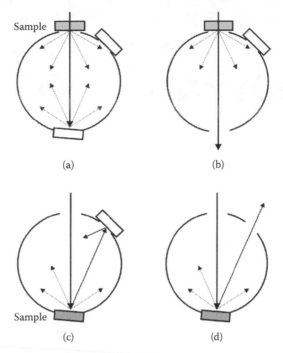

**FIGURE 3.14**    The method of measuring (a) $T_{tot}$, (b) $T_{dif}$, (c) $R_{tot}$, and (d) $R_{dif}$ of the samples, using an integrating sphere.

The internal surface of the integrating sphere is covered with a highly reflective and diffusive white material (e.g., Teflon® or Spectralon®) to reflect the light efficiently and diffuse it uniformly over the entire internal surface of the sphere, including the spot where the photodetector is placed. By considering the uniform distribution of light and by knowing the surface area of the sphere, the entire intensity of light collected inside the sphere can be determined from the intensity reaching the detector. Integrating spheres usually have more openings to control light paths and to allow different sample placements. An example of a sphere with three basic openings is shown in Figure 3.15.

The opening (1) at the front side of the sphere enables the entrance of a light beam coming from a monochromator, and the opening (2) at the back side is used for reflectance measurements or for elimination of the specular transmitted beam when the transmittance is measured. The opening denoted (3) is used for elimination of the specular beam of reflected light when measuring diffused reflectance.

For measuring $T_{tot}$ we put the semi-transparent sample with textured surface in front of the opening (1) where the light enters the sphere (Figure 3.14a). All other openings are closed with removable mirrors covered with the same material as the inner surface of the sphere. In this way all the light that is transmitted through the sample (scattered and specular component) is captured

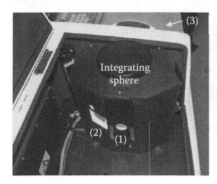

**FIGURE 3.15**   The integrating sphere of a PerkinElmer Lambda 950 spectrometer. (1) The opening for the light entrance, (2) the opening at the back side used for diffused transmittance and in reflectance measurement, and (3) the opening used for diffused reflectance measurement.

in the sphere and detected by the photodetector. To measure only diffused transmittance, $T_{dif}$, we remove the mirror at the opening (2) at the back side of the sphere; thus, specular transmitted light escapes the sphere and is not detected (Figure 3.14b). For reflectance measurements we place the sample at the back side of the opening (2) instead of the mirror used in total transmittance measurements. $R_{tot}$ is measured if the mirror is present at the opening (3) (Figure 3.14c), and $R_{dif}$ is detected if the mirror is removed, since the reflected specular beam can escape the sphere (Figure 3.14d).

Examples of measurements on different samples used as substrates in thin-film solar cell technology are presented in Section 3.5.

Are TIS measurements easy to perform? If reflectances and transmittances are required to be measured very precisely with an integrating sphere, we have to be aware that some errors occur as a result of a small amount of light that escapes out of the sphere throughout the front opening (1) where the light originally enters. This may lead to somewhat decreased values of reflectance and transmittance. However, when calculating the haze parameter, the diffused components are divided by the total ones, and since the decrease is present in both cases it is partially suppressed.

The second issue is that in TIS measurements the samples are usually surrounded by air. If only one external surface of the sample is textured, we can attribute the measured scattering properties to the air/sample (sample/air) interface. However, reflections at another surface or internal interfaces may affect the measurements. For measuring of $T_{tot}$ of TCO samples (glass/TCO) often an index matching liquid (IML) is applied in the system, as shown in Figure 3.16.

In this case scattering level at textured TCO surface is suppressed significantly since the refractive index of the IML is supposed to be close to the one of the TCO layer. Measured total transmittance in this system is usually higher

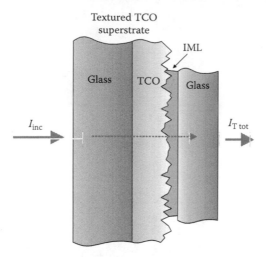

**FIGURE 3.16**  Total transmittance measurements of a textured TCO superstrate by means of an index matching liquid (IML).

than the one measured directly in the air surrounding, since there is much less reflected scattered light at the TCO/IML interface. This scattered light may escape the sample at the front side or at the edges of the glass substrate, decreasing the measured transmittance.

### 3.2.6.2  ANGULAR RESOLVED SCATTERING MEASUREMENTS

The angular resolved scattering (ARS) measurement system enables us to detect an angular dependency of scattered light at a textured surface. Why is this angular characteristic important for a solar cell device? The light beams that are scattered into larger scattering angles experience longer optical paths when propagating throughout a layer. In addition, if the angles of propagation inside a layer are large enough, the condition of total reflection can be fulfilled at the surrounding interfaces, leading to significant improvement in light trapping in the device.

In principle, an ARS system consists of a rotary stage with a photodetector that rotates around the illuminated sample and measures the intensity of scattered light at different (discrete) angles. A schematic layout of the system and a photo of one of the possible realizations are shown in Figures 3.17 and 3.18.

As a source of illumination, gas lasers or collimated laser diodes are used to obtain a sufficiently large (measurable) intensity of scattered light. Laser sources of different wavelengths can be employed to detect the wavelength dependency of *ADF*s. Solutions with a broadband light source in combination with a monochromator are also possible (integrated system inside the spectrometer) (Jäger et al. 2011), although we can expect more noise in the measuring signal due to a relatively low intensity of incident and consequently of scattered light.

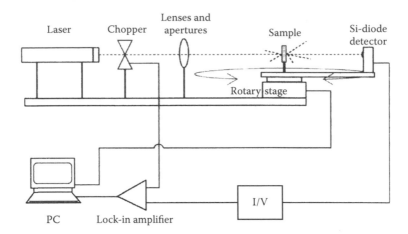

**FIGURE 3.17**   A schematic representation of an ARS system.

In any case, we have to apply a lock-in technique to minimize the effects of background illumination and to reduce the noise level. Thus, the light beam is chopped either mechanically by a chopper wheel or electronically in the case of laser diodes. The chopped laser beam can be sharpened by additional lenses and apertures before impinging the sample.

**FIGURE 3.18**   A photo of an ARS system (University of Ljubljana).

A silicon (or germanium) photodetector is used to detect the intensity of the visible or infrared light signal at a certain scattering angle of a measuring plane. The light signal is transformed in the photodetector to electrical current. The photodetector unit is mounted on a rotating arm that is precisely rotated in a measuring plane by means of a motorized system (like stepper motor). The detector arm should be able to move in a full-angle range from 0 degrees to 360 degrees, enabling the measurement of both reflected and transmitted scattered light of the sample.

The electrical current of the photodetector, which corresponds to the measured chopped light intensity, is transferred and amplified to a voltage signal by means of a sensitive current-to-voltage amplifier. The amplifier is then connected to the lock-in amplifier where the final voltage signal can be obtained using the reference chopper signal. The chopper frequency is typically from 10 Hz to 1 kHz, which is far below the limitation from the detector's response time or the amplifier's bandwidth. The measured data are transferred to the computer. The values of the output voltage signal are linearly dependent on the scattered light intensity at the corresponding scattering angles. We usually normalize the measured values such that the value corresponding to the highest diffused intensity (usually the one that is detected near the specular beam) is set to unity. In this way we define a normalized *ADF*, which ranges from 0 to 1. The normalized *ADF*, to which we refer in all further investigations, gives information about the relative distribution of light intensity in different directions, whereas the absolute scattering level with respect to the specular light is defined by the haze parameter. We present examples of *ADF*s corresponding to realistic substrates in Section 3.5.

The *ADF* is usually measured for the sample in air surrounding. However, there are solutions in which we can get information also about directional scattering in other media (like water or IMLs). Two possible configurations of such measurements are presented in Figure 3.19 for the case of transmitted light. In Figure 3.19a, the sample with a textured surface is located centrally inside a glass sphere filled with a liquid. The liquid used (usually water) represents the surrounding medium of the sample. At the internal and external side of the rounded transparent walls of the sphere, one expects certain reflections that may affect the measurements. However, since the incident angles of scattered light beams are perpendicular to the walls and since no high difference in refractive indexes of the liquid, sphere, and outside air is typically present, these reflections are usually small and can be ignored. Another example of the measurement of *ADF* in a medium other than air is presented in Figure 3.19b, where a small half-sphere made of a certain low-absorbing material (such as glass or sapphire) is attached to the textured surface of the sample by an IML, which should have approximately the same refractive index as that of the material of the hemisphere. The advantage of this second method is that less liquid is required (usually only a drop of IML) and that small hemispheres of different materials are commercially available.

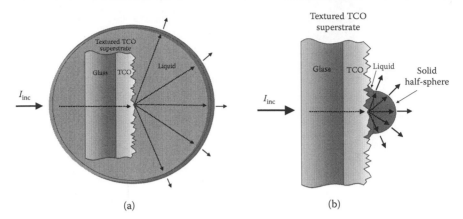

(a)                                                    (b)

**FIGURE 3.19**    $ADF_T$ measurements in a surrounding medium other than air. (a) A solution with a sphere filled by a liquid. (b) A solution with a small half-sphere attached by index matching liquid to the sample surface.

### 3.2.6.3   TRANSFORMATION OF ANGULAR DISTRIBUTION FUNCTIONS

In this subsection we raise and answer a question whether the *ADF* that we measure in a plane with an ARS system is really the *ADF* that we should consider in optical simulations. When measuring the $ADF_T$, or $ADF_R$ in a plane around the sample, the scattering in this plane only, and not in the entire 3-D space, is detected. However, in optical simulations, 1-D or more dimensional, we have to take into account the intensity of entire scattered light and not just the one in the measuring plane. Since the roughness morphology of the randomly textured substrates used in thin-film solar cell technology is normally isotropic, the light scattering is assumed to be spherically symmetrical around the point of illumination. Therefore, it turns out to be sufficient to detect the intensity of scattered light in only one cross-sectional measuring plane, but we have to apply a transformation to include all 3-D scattering into solid angles.

Because the same transformation is applicable to the case of reflected and transmitted scattered light, we explain it for the case of transmitted light only. The transmission light-scattering process at a semi-transparent sample with a textured surface is illustrated in Figures 3.20 and 3.21. We define a differential segment of the hemisphere in which the light is scattered under the same angle φ. Its area, $\Delta A(\varphi)$, we determine by Equation (3.11):

$$\Delta A(\varphi) = 2 \cdot \pi \cdot R \cdot h \tag{3.11}$$

where $R$ represents the distance from the sample (scattering point) to the detector and $h$ is the thickness of the hemisphere segment as shown in Figure 3.21.

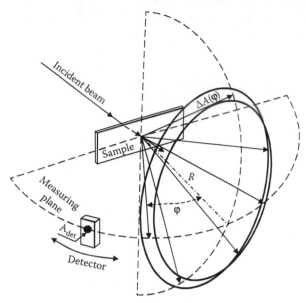

**FIGURE 3.20**   Light scattering at a textured surface in a 3-D space. Originally, in the ARS measurements only the intensities of the rays scattered in a measuring plane are detected.

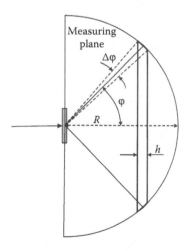

**FIGURE 3.21**   Top view of the 3-D scattering problem. The segment of the hemisphere with the thickness of $h$ is defined.

Considering the given geometry, we can determine $h$ using Equation (3.12):

$$h = 2 \cdot R \left( \cos\varphi - \cos\left(\varphi + \frac{\Delta\varphi}{2}\right) \right) \tag{3.12}$$

where $\varphi$ is the scattering angle and $\Delta\varphi$ is the discrete angular step in our measurements. If $\Delta\varphi$ is small we can consider that the power density, $J$, of the scattered light is constant all over the area $\Delta A(\varphi)$. This power density is detected in the point of the measuring plane by the photodetector. The voltage signal $V_{det}$ at the output of the lock-in amplifier is proportional to the power density $J$ as given in Equation (3.13):

$$J = \frac{k \cdot V_{det}}{A_{det}} \tag{3.13}$$

where $A_{det}$ is the area of the detector, $V_{det}$ is its output voltage signal, and $k$ is the scaling factor that depends on the sensitivity of the detector and on the amplification level of the amplifiers. The light intensity denoted by $I_{3\text{-}D}(\varphi)$ that is scattered in the area $\Delta A(\varphi)$ can be determined by Equation (3.14).

$$I_{3\text{-}D}(\varphi) = J \cdot \Delta A(\varphi) = k \cdot \frac{V_{det}}{A_{det}} \cdot 4\pi R^2 \cdot \left( \cos\varphi - \cos\left(\varphi + \frac{\Delta\varphi}{2}\right) \right) \tag{3.14}$$

After normalization of $I_{3\text{-}D}(\varphi)$ with the sum of all contributions at different discrete $\varphi$, we obtain the final angular distribution function, $ADF_{3\text{-}D,1\text{-}D}$, which includes light scattering in a solid angle of the 3-D space and which we should consider in optical simulations.

Considering the trigonometry rules (Bronshtein et al. 2007) we can further express the term $(\cos(\varphi)\text{-}\cos(\varphi + \Delta\varphi/2))$ from Equation (3.14) as given in Equation (3.15).

$$\cos\varphi - \cos\left(\varphi + \frac{\Delta\varphi}{2}\right) =$$

$$= \cos\varphi - \cos\varphi \cdot \cos\frac{\Delta\varphi}{2} + \sin\varphi \cdot \sin\frac{\Delta\varphi}{2} \approx \tag{3.15}$$

$$\approx \sin\frac{\Delta\varphi}{2} \cdot \sin\varphi; \quad \cos\frac{\Delta\varphi}{2} \approx 1 \quad \textit{if } \Delta\varphi \textit{ is small}$$

Since in the ARS measurements $\Delta\varphi$ is typically small (e.g., $\Delta\varphi = 1$ degree) and, thus, $\cos(1°/2) = 0.9996 \simeq 1$, we can use the final approximation of the term $(\cos(\varphi) - \cos(\varphi + \Delta\varphi/2)) \simeq \sin(\Delta\varphi/2) \cdot \sin(\varphi)$, resulting in a simplified Equation (3.16).

$$I_{3\text{-D}}(\varphi) = k \cdot \frac{V_{det}}{A_{det}} \cdot 4\pi R^2 \cdot \sin\frac{\Delta\varphi}{2} \cdot \sin\varphi \qquad (3.16)$$

If we combine all the constant factors of the measurement ($k$, $A_{det}$, $R$, $\Delta\varphi$) in one constant, $c$, Equation (3.16) can be further simplified, as in Equation (3.17).

$$I_{3\text{-D}}(\varphi) = c \cdot V_{det} \cdot \sin\varphi \qquad (3.17)$$

Thus, the normalized *ADF* in 3-D space (denoted by $ADF_{3\text{-D, 1-D}}$) can be calculated from the *ADF* measured in a plane as given in Equation (3.18)

$$ADF_{3\text{-D,1-D}}(\varphi) = c_n \cdot ADF(\varphi) \cdot \sin\varphi \qquad (3.18)$$

where $c_n$ is a constant including $c$ and the normalization of $ADF_{3\text{-D, 1-D}}$.

The transformation of *ADF* into $ADF_{3\text{-D, 1-D}}$ is illustrated for an example of Lambertian (cosine) *ADF* function in Figure 3.22 by the polar plot and in Figure 3.23 by the Cartesian plot.

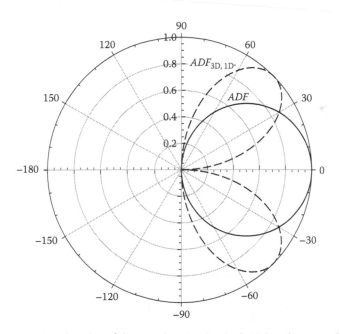

**FIGURE 3.22**    A polar plot of the Lambertian (cosine) *ADF* and its transformation into $ADF_{3\text{-D, 1-D}}$, which includes light scattering in all three dimensions, assuming the spherical symmetry of scattered light.

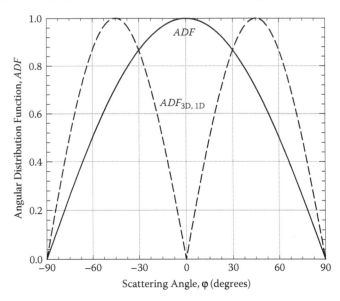

**FIGURE 3.23**   Lambertian (cosine) *ADF* and its transformation into $ADF_{3\text{-}D,\ 1\text{-}D}$ represented in the Cartesian plot.

If we apply the illumination under a non-perpendicular incident angle, the transformation of a 3-D scattering is more complicated. How do we deal with it? Here we should consider spherical symmetry in the incident illumination, resulting in a cone-like illumination, defined by the incident angle $\varphi_{inc}$, as shown in Figure 3.24.

In the ARS measurements we typically apply only one laser beam under a certain angle of incidence, forming one narrow part of the cone-like illumination. However, even from this kind of measurement we can reconstruct the *ADF* of the cone-like illumination. The cone-like illumination is suitable for 1-D modeling where cylindrical symmetry has to be assumed.

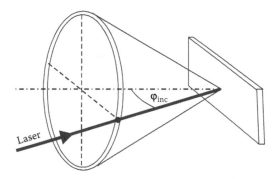

**FIGURE 3.24**   Illumination of the sample with a laser beam as a small discrete part of a cone-like illumination under incident angle $\varphi_{inc}$.

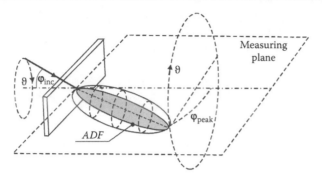

**FIGURE 3.25**   3-D scattering problem for the case of non-perpendicular illumination.

The *ADF* detected in a measuring plane represents only a cross-section of a 3-D angular distribution, as explained earlier. For the case of non-perpendicular illumination, the situation that corresponds only to one beam illumination (as a part of cone-like illumination) is sketched in Figure 3.25, whereas the principle of how to consider the cone-like simulation is illustrated in Figure 3.26. To determine the 3-D scattering from the measurements in the cross-sectional plane in this case we assume a cylindrical symmetry around the middle *ADF* axis located at $\varphi_{peak}$ (Figure 3.25). If we rotate the incident beam around the incident axis (dash-dot line) following the angle $\vartheta$, we can imitate the entire cone-like illumination. By rotation of the incident beam, the 3-D shape of the *ADF* also rotates. Based on this we have to calculate the new cross-sections corresponding to each discrete angle $\vartheta$ as illustrated in Figure 3.26. Then we have

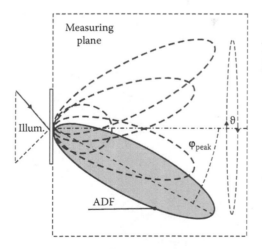

**FIGURE 3.26**   The intersections of the 3-D scattering body shapes with the measuring plane. The shapes correspond to different discrete segments of the cone-like illumination.

to sum up all the cross-sectional contributions, resulting in the common *ADF* of the cone-like illumination, still corresponding to a plane of measurements. Thus, we have to perform the transformation of the common *ADF* to $ADF_{\text{3-D, 1-D}}$ as in the case of perpendicular illumination.

As an example of the procedure explained, in Figures 3.27 and 3.28 we present the common angular distribution function, $ADF_{\text{cone ill}}$, of a cone-like illumination for realistic *ADF* of a ZnO:Al surface-textured substrate (see Section 3.5.2). The illumination angle is chosen to be $\varphi_{\text{inc}} = 60$ degrees (plotted by circles). The approximation of the $ADF_{\text{cone ill}}$ with a more simple function $ADF_{\text{appr}}$ is also shown.

In optical simulations we have to determine the common *ADF* for more incident angles $\varphi_{\text{inc}}$ of the illumination, since at the interfaces of the solar cell structures light impinges under various angles because of a multiple scattering process. In the case of complex *ADF* functions it is recommended to use some simplifications in the procedure of calculations of common *ADF* for different incident angles. For example, we can calculate the common *ADFs* only for a few selected $\varphi_{\text{inc}}$ and then use an interpolation. Second, the resulting *ADFs* we approximate with simple analytical functions (as shown in Figures 3.27 and 3.28). These functions can then be incorporated in optical simulations.

In principle, from the results of ARS measurements we can determine the total scattered (diffused) light intensity, $I_{\text{dif}}$, by integrating the contributions $I_{\text{3-D}}(\varphi)$ for

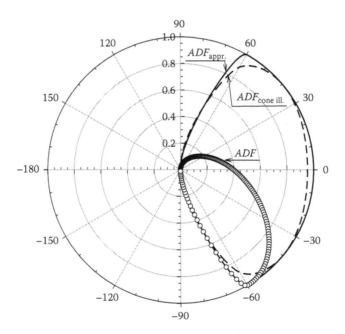

**FIGURE 3.27**  An example of a common $ADF_{\text{cone ill}}$ for the cone-like illumination determined from the *ADF* that corresponds to one laser-beam illumination. A possible approximation of the $ADF_{\text{cone ill}}$ is also added in the polar plot ($ADF_{\text{approx}}$).

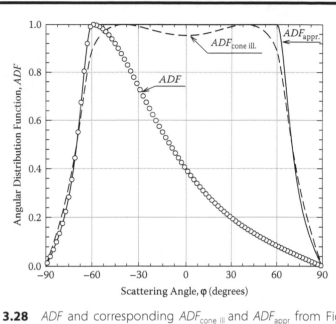

**FIGURE 3.28**    *ADF* and corresponding $ADF_{cone\ ill}$ and $ADF_{appr}$ from Figure 3.27 represented in a Cartesian plot.

all the angles $\varphi$. To determine haze parameters from the ARS measurements, the specular component $I_{spec}$ has to be measured in order to obtain the total intensity $I_{tot}$. The determination of the specular component by means of the ARS measurements is not a trivial issue. Why? The main problem is that approaching the specular direction the light intensity is changing rapidly and with a detector of a finite area the border between specular and scattered light cannot be defined exactly. This uncertainty may lead to an error in haze determination of a factor of up to 2 or more as indicated by practical observations. Therefore, TIS measurements are used to determine haze parameters and the ARS measurements are supposed to be used to determine *ADF*s primarily.

### 3.2.6.4   SCATTERING PARAMETERS OF INTERNAL INTERFACES

In optical simulations of thin-film solar cells we have to define scattering parameters of the interfaces inside the solar cell structure (internal interfaces). We usually carry out the measurements of scattering parameters by TIS and ARS on the substrate/superstrate samples with a textured surface facing air or some other surrounding medium (such as glass or sapphire, applying IML between the sample and the medium). In solar cells, textured interfaces are formed by other materials. Scattering parameters inside the solar cells cannot be measured directly; thus, a model is needed to link the textured interface morphology and optical properties of adjacent layers with the scattering behavior inside the structure. For the random morphologies that are used in thin-film solar cells, the exact correlation between the morphology and the scattering properties can be obtained by means of applying rigorous methods of solving Maxwell equations in a numerical way (such as presented in Chapters 4 and 6 for periodic textures). However, in a complex optical system with multiple layers and randomly textured interfaces, such as thin-film solar cells are, this action requires a 3-D approach with many discretization points to describe random interface morphology. Time-consuming simulations with powerful computers are often required if following this approach. On the other hand, approximations based on scalar scattering theory present a simpler and quite reliable solution (Zeman et al. 2000; Krč et al. 2003; Jäger and Zeman 2009), although they are not fully applicable to the random morphologies of the substrates used. Scattering parameters obtained externally by TIS and ARS serve as a calibration of the method applied. A nice comparison between the approach of rigorous 3-D modeling to determine the scattering parameters and an approach based on scalar scattering theory is presented in Rockstuhl et al. (2011), where the authors indicate that the simple scalar theory approaches work very well in many cases.

The scalar scattering theory is widely employed in optical modeling of thin-film solar cells to determine the specular or diffused part of reflected and transmitted light at a textured interface (Stiebig et al. 2000; Zeman et al. 2000; Krč et al. 2002; Springer, Poruba, and Vaneček 2004). For determination of haze parameters according to Beckmann and Spizzichino (1987), there are two critical validity conditions of this theory which are, however, often exceeded when applying it to the realistic nano-textured interfaces of thin-film solar cells:

- The lateral correlation length, $L$, of the interface morphology has to be much larger than the wavelength of light in the medium ($L \gg \lambda$). This also means that the radius of the curvatures of the interface irregularities is much larger than the wavelength of light. This is required for the validity of the Kirchoff approximation (Kong 1990) used in the theory. In this case also the mutual interaction of the irregularities (e.g., shadowing and multiple scattering) could be neglected.
- The height distribution of the irregularities forming the interface texture is assumed to be a Gaussian function.

Between the two, the first condition is more critical to be fulfilled, since typically the features of the textures at the interfaces of thin-film solar cells are in the range of light wavelengths and often have sharp edges. Substrate textures with pyramid-like features have the height distribution close to Gaussian. Other types of morphologies such as crater-like, which is an example of texturization obtained by wet etching of the ZnO TCO layer, may deviate from Gaussian distribution (Kluth et al. 1999). We can compensate these violations of the two conditions at least partially and usually successfully by introducing the calibration functions in the original equations of the scalar scattering theory used for haze determination as shown later.

### 3.2.6.5   DETERMINATION OF HAZE PARAMETER BY USING SCALAR SCATTERING THEORY

Following the theory we define the haze parameters for reflected and transmitted light (Carniglia 1979; Beckmann and Spizzichino 1987) as

$$H_R = 1 - \exp\left[ -\left( \frac{4\pi \cdot \sigma_{rms} \cdot n_l \cdot \cos\varphi_{inc}}{\lambda} \right)^2 \right] \tag{3.19}$$

$$H_T = 1 - \exp\left[ -\left( \frac{2\pi \cdot \sigma_{rms} \cdot |n_l \cos\varphi_{inc} - n_r \cos\varphi_T|}{\lambda} \right)^2 \right] \tag{3.20}$$

where $\sigma_{rms}$ is the vertical *rms* roughness parameter of the textured interface, $n_l$ and $n_r$ are real parts of wavelength-dependent complex refractive indexes of the left and right layers forming the interface, $\lambda$ is light wavelength in free space, and $\varphi_{inc}$ and $\varphi_T$ are the incident angle and the angle of transmitted specular beam, respectively. The equations present the required link between interface morphology described by only one representative parameter, $\sigma_{rms}$, optical properties of adjacent layers ($n_l$, $n_r$) and the properties of light ($\lambda$, $\varphi_{inc}$, $\varphi_T$). Equations (3.19) and (3.20) show that by increasing the $\sigma_{rms}$, the haze increases. As a function of $\lambda$, the haze parameters decrease quasi exponentially according to the equations, if going to longer wavelengths.

The proposed equations of scalar scattering theory present a very efficient, fast, and simple way to link the morphology, optical properties, and haze parameter of an interface even in a multilayer system model. The general quasi-exponential decay as a function of wavelength can be observed also in the measurements of textured substrates used for thin-film solar cells. However, before we apply them to internal interfaces, the calculations have to be compared and calibrated with the measurements on the textured substrates used. As shown later, the exact fit with the measurements usually cannot be obtained through use of the original form of the equations. Therefore, we apply calibration functions and factors to the original equations, and this leads to a better approximation of the measured results. Applying these calibrations we implicitly compensate the violations of morphological restrictions of the theory. Since in the solar cells the surface morphology of the substrate is transferred to other interfaces with some decrease in $\sigma_{rms}$ only (lateral features of the morphology remain more or less the same), we then assumed these calibrations lead to a better approximation of haze parameters of internal interfaces as well.

In particular, we introduce in the equations for haze two calibration functions $c_T$ and $c_R$ [Equations (3.21) and (3.22)], which can be dependent on light wavelength and the $\sigma_{rms}$ parameter in general.

$$H_R = 1 - \exp\left[ -\left( \frac{4\pi \cdot c_R(\lambda, \sigma_{rms}) \cdot \sigma_{rms} \cdot n_l \cdot \cos\varphi_{inc}}{\lambda} \right)^2 \right] \quad (3.21)$$

$$H_T = 1 - \exp\left[ -\left( \frac{2\pi \cdot c_T(\lambda, \sigma_{rms}) \cdot \sigma_{rms} \cdot |n_l \cos\varphi_{inc} - n_r \cos\varphi_T|}{\lambda} \right)^a \right] \quad (3.22)$$

Following the original equations, $c_R$ and $c_T$ are unity. In practice, they can vary between 0.5 and 1.5 for the realistic textures of interfaces within thin film solar cells. Another calibration factor that we can introduce in the equation of $H_T$ is the power factor $a$. This factor changes the slope of exponential decay of $H_T$ as a function of wavelength. Originally its value is set to 2, but practically it may vary from 1.8 to 3 (Stiebig et al. 2000). The calibrations of haze parameters for selected examples of realistic substrates are presented in Section 3.5.

### 3.2.6.6    APPLICATION OF ANGULAR DISTRIBUTION FUNCTIONS TO INTERNAL INTERFACES

As we can take the equations of scalar scattering theory as a basis to determine the haze parameters of internal interfaces in optical simulations of thin-film solar cells with randomly nano-textured interfaces, it has been only recently reported and investigated how to apply scalar scattering theory to the determination of internal *ADF*s in thin-film solar cells (Jäger and Zeman 2009; Dominé et al. 2010). These approaches are based on first Born approximation and Fourier transform of the surface texture in order to estimate the *ADF* of a particular interface. Other approaches based on geometrical optics applied to nano-textures have been reported as well. In this case, we also have to be aware that the validity conditions of geometrical optics (texturization features should have been much larger than the wavelengths of light) are violated. The exact method of determining the scattering properties of an interface would be based on rigorous solving of Maxwell's equations as already mentioned.

Despite the mentioned restrictions, we apply the approximations to nano-textured surfaces of realistic substrates, being aware of some violations of validity conditions. In the following we present an example of a simplified geometrical optics approach to estimate the *ADF* of nano-textured interfaces. As the input, interface morphology obtained by AFM and complex refractive indexes, forming the interface, are taken into account. In calculations unpolarized light was applied perpendicularly to the surface plane.

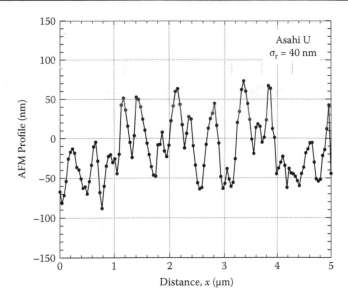

**FIGURE 3.29**    AFM profile of the textured surface of the Asahi U-type $SnO_2$:F superstrate. The dots correspond to the measured values.

We analyze two types of surface textured superstrates: Asahi U-type TCO on glass with naturally textured $SnO_2$:F TCO film and chemically etched ZnO:Al on glass superstrate with textured ZnO:Al surface. In Figures 3.29 and 3.30 the representative AFM scans of the $SnO_2$:F surface with $\sigma_{rms} = 40$ nm and ZnO:Al surface with $\sigma_{rms} = 60$ nm are shown, respectively. Since the morphology of the analyzed substrates is isotropic, we consider only the AFM scan in only one (arbitrary) direction with respect to the sample orientation in the analysis. In the scans $128 \times 128 = 16,384$, we used equidistant AFM datapoints obtained at the area $5 \times 5$ $\mu m^2$ on the sample. The distance between two points was $5$ $\mu m/128 \simeq 40$ nm. The $16,384 \times 40$ nm $\simeq 640$ $\mu m$ long scan was obtained by combining 128 horizontal scans with 128 points in one row. In the calculation we apply discrete light beams, representing the incident illumination, to each segment determined by linear connection of the two neighbor AFM points. Microscopically, each segment can be taken as a differential flat interface with a certain inclination with respect to the incident light beam. The incident angles can be defined and the reflected and transmitted light beams determined (incidence of incoherent light at a flat surface). We sum up all the intensities of the reflected and transmitted beams that propagate in the same discrete directions. From these common intensities assigned to the corresponding angles, the angular distribution $ADF_R$ and $ADF_T$ can be constructed.

The results of the calculated $ADF_T$s in comparison to the measured ones are plotted in Figures 3.31 and 3.32.

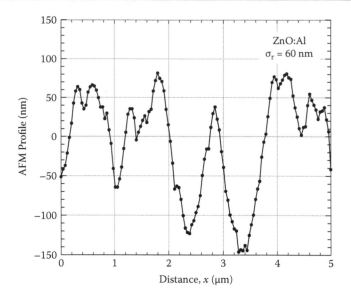

**FIGURE 3.30**   AFM profile of the textured surface of a glass/ZnO:Al superstrate. The dots correspond to the measured values.

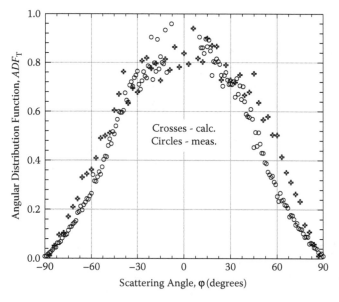

**FIGURE 3.31**   Measured and estimated (calculated) $ADF_T$ of the Asahi U-type $SnO_2$:F superstrate for perpendicular incidence of light.

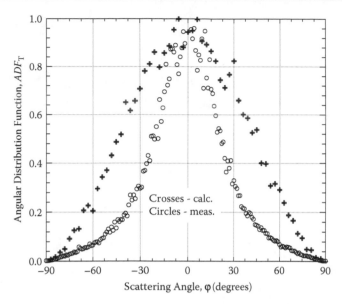

**FIGURE 3.32** Measured and estimated (calculated) $ADF_T$ of the glass/ZnO:Al superstrate for perpendicular incidence of light.

Figures 3.31 and 3.32 reveal that through using this simple approach, only a rough approximation of the $ADF_T$ can be obtained, particularly in the case of ZnO:Al surface, which exhibits narrower $ADF_T$ in reality. However, comparing the trends in measured and calculated $ADF_T$s for the $SnO_2$:F/air and ZnO:Al/air interfaces (Figure 3.33) we can observe the same general behavior of broader $ADF_T$ function also in calculations (Figure 3.34).

To improve the presented method of geometrical optics approach, we have to determine for each segment the possible escaping cone of the transmitted (reflected) beam as shown in Figure 3.35.

In the first step we can assume that only the beam that is propagating within the escaping cone can contribute to the distribution function. Figures 3.31 and 3.32 indicate that the calculated $ADF_T$s are too broad, compared to the measured ones. This can indicate that the scattering into larger angles is overestimated. Considering the escaping cones may lead to a better approximation.

Further on, secondary scattering of the light beams exceeding the escaping cone can be included in the calculation. We have to determine the segment on which the exceeding light beam falls and define the new incident angle. Again, the corresponding reflected and transmitted beams are defined. Only one or both, reflected and transmitted beams, we trace further.

In 1-D semi-coherent optical simulations often as a first approach the $ADF$s as measured in the air are imported into the model and applied to internal interfaces, due to simplicity reasons. Snell's law could be applied as an approximation to redefine the angles of scattered light according to refractive indexes

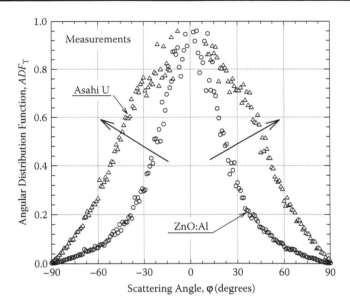

**FIGURE 3.33** The difference in the broadness of the measured *ADF*$_T$ of the Asahi U-type superstrate and the glass/ZnO:Al superstrate, indicated by the arrows.

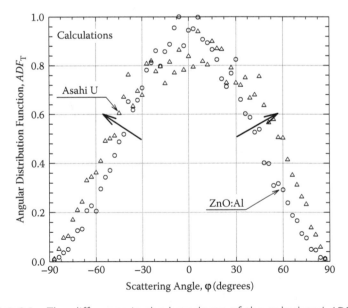

**FIGURE 3.34** The difference in the broadness of the calculated *ADF*$_T$ of the Asahi U-type superstrate and the glass/ZnO:Al superstrate, indicated by the arrows.

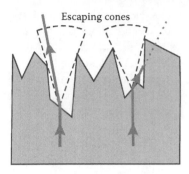

**FIGURE 3.35**   Introduction of escaping cones defining which transmitted beams can contribute to the calculated $ADF_T$.

of the layers. However, more accurate approaches of defining the $ADF$ of internal interfaces are preferable.

### 3.2.7   OTHER OPTICAL PROPERTIES OF THE INTERFACES

Besides light scattering at nano-textured interfaces, the level of total reflection and transmission of the interface can be affected by antireflection properties of the texturization. This effect becomes pronounced if sub-wavelength textures are present at the interface (textures with smaller vertical and lateral dimensions than those of light wavelength). Antireflection at textures can be assigned to a smoother transition from one material to another, following the light propagation, compared to the flat interface with an abrupt change in refractive index at the interface plane.

For sub-wavelength textures we often model the antireflecting effect by effective media approach (EMA), using the Lorenz-Lorenz, Maxwell Garnett, or Bruggeman approach (Ohlidal, Navratl, and Ohlidal 1995; Choy 1999). Basically, with EMA we can determine the decrease in total reflectance due to surface texturization, whereas for determination of light-scattering properties we have to apply other theories discussed in previous sections. In EMA we present the textured interface region with one or more fictitious sub-layers with flat interfaces whose optical properties are represented by a mixture of the properties of the materials forming the interface. The mixture fractions ($f$ and $1 - f$) we determine by actual volume fraction of the two materials present in the volume that is occupied by the EMA sub-layer. Following the Bruggeman approach, optical properties (dielectric functions) of an EMA sub-layer at a textured interface formed by two materials can be determined via Equation (3.23):

$$\frac{\varepsilon_1 - \varepsilon_{EMA}}{\varepsilon_1 + 2\varepsilon_{EMA}}(1 - f) + \frac{\varepsilon_2 - \varepsilon_{EMA}}{\varepsilon_2 + 2\varepsilon_{EMA}}f = 0 \qquad (3.23)$$

where $\varepsilon_1$ is the complex dielectric function of the first material with its volume fraction $(1 - f)$ and $\varepsilon_2$ is the complex dielectric function of the second material with its volume fraction $(f)$ in the region of the EMA layer. The complex dielectric function of the EMA sub-layer is described by $\varepsilon_{EMA}$.

We can introduce EMA sub-layers into simulated structure as new layers located between the existing layers, forming the textured interface. However, in addition to the antireflecting effect, we should preserve also the light-scattering properties of the interface, if present. Thus, we should apply at least one textured interface with the required scattering properties into the EMA stack.

In the Sun*Shine* optical simulator, for example, another option was introduced, which enables us to correct the existing reflectivity function of an interface calculated from the complex refractive indexes. We correct the reflectance for the electric field as given in Equation (3.24):

$$r_{corr} = \sqrt{c} \cdot r_{orig} \tag{3.24}$$

where $c$ is the correction factor and $r_{orig}$ is the original reflectance calculated from complex refractive indexes (Fresnel coefficient). We apply the square root to the factor $c$, since it basically applies as a correction of reflectances defined on the intensity level.

We can calculate then the corrected transmittance $t_{corr}$ from

$$r_{corr} = 1 + t_{corr} \tag{3.25}$$

ensuring the energy conservation at the interface.

On the level of light intensity, the reflectance and transmittance correction is applied as given in Equations (3.26) and (3.27).

$$R_{corr} = c \cdot R_{orig} \tag{3.26}$$

$$T_{corr} = 1 - R_{corr} \tag{3.27}$$

In general the parameter $c$ can be a function of light wavelength and incident angle. It can even be determined separately for left- and right-side illumination in the simulator. By introducing such reflection correction determined either experimentally or theoretically (such as EMA), we can consider the antireflecting effect related to the surface textures without introducing any fictitious layers in the structure in simulations. This option presents a useful tool to study and manipulate the reflectivity properties of a particular interface in solar cell structure artificially. However, possible realizations of such reflectance manipulations must then be found for realistic structures. In general, one of the solutions offers photonic crystal structures with wavelength selective reflectivity properties. These advanced structures are addressed in Section 5.2.5 in Chapter 5.

Besides textured interfaces the interface reflectance can be decreased also as a result of formation of thin interfacial films between two layers or at a layer surface. In this case we have to introduce special buffer layers with average optical properties of the two layers in optical simulations. Two examples where such interfacial layers were reported are related to the back contact in solar cells. One is at the n-a-Si:H/Ag interface in amorphous silicon solar cells (Stiebig et al. 1994) and the other is at the CIGS/Mo interface in chalcopyrite solar cells (Krč et al. 2006). In the n-aSi:H/Ag interface, a thin layer (1.5 nm) with the optical properties of Al can be introduced as a buffer layer in simulations of amorphous silicon solar cells without a back TCO layer (Stiebig et al. 1994). However, instead of buffers we can use a proper reflectance correction function in simulations, if the simulator allows it.

## 3.3    OUTPUT PARAMETERS

The main output parameters of optical simulations are

- reflectance and transmittance of the structure,
- absorptance in individual layers,
- light intensity profile in the structure, and
- charge carrier generation rate profile.

We define total, diffused, and specular reflectance and transmittance of the structure as

$$R_{\text{tot}} = \frac{I_{\text{R tot}}}{I_{\text{inc}}} \qquad R_{\text{dif}} = \frac{I_{\text{R dif}}}{I_{\text{inc}}} \qquad R_{\text{spec}} = \frac{I_{\text{R spec}}}{I_{\text{inc}}} \qquad (3.28)$$

$$T_{\text{tot}} = \frac{I_{\text{T tot}}}{I_{\text{inc}}} \qquad T_{\text{dif}} = \frac{I_{\text{T dif}}}{I_{\text{inc}}} \qquad T_{\text{spec}} = \frac{I_{\text{T spec}}}{I_{\text{inc}}} \qquad (3.29)$$

where $I_{\text{inc}}$, $I_{\text{R tot}}$, $I_{\text{R spec}}$, and $I_{\text{R dif}}$ are the intensities of the incident, reflected total, reflected specular, and reflected diffused light, respectively. For transmitted components, subscript "R" is exchanged with subscript "T."

The total absorptance in the individual layer of the structure can be defined as

$$A_{\text{tot}} = \frac{I_{\text{tot in}} - I_{\text{tot out}}}{I_{\text{inc}}} \qquad (3.30)$$

where $I_{\text{tot in}}$ and $I_{\text{tot out}}$ are the total intensities of light that enter and the one that leaves out the layer, respectively. Specular or diffused components can be used in the denominator to determine specular and diffused absorptances.

The light intensity profile in the structure is presented by a combination of forward- and backward-going intensities, including their interactions.

We can determine the charge carrier generation rate profile $G_L(x,\lambda)$ in the semiconductor layers as

$$G_L(x,\lambda) = -\frac{\lambda}{h \cdot c} \cdot \frac{dI_{tot}(x,\lambda)}{dx} \tag{3.31}$$

where $h$ is the Planck constant ($h = 6.625 \cdot 10^{-34}$ Js), $c$ represents the velocity of light ($c = 3 \cdot 10^8$ m/s), and $x$ is the spatial component. To obtain a total charge carrier generation profile, $G_L(x)$, of the entire solar spectrum, we have to sum up all of the discrete wavelength contributions $G_L(x,\lambda_i)$ as shown by Equation (3.32)

$$G_L(x) = \sum_{i=1}^{i=K} G_L(x,\lambda_i) \tag{3.32}$$

where $K$ is the total number of discrete components in the incident light spectrum. The $G_L(x,\lambda)$ profile presents a basis for further electrical analysis of the multilayer structure.

In thin-film solar cell simulations we often combine optical simulations with a simplified (but justified) electrical analysis to determine the output parameters of the solar cells such as external quantum efficiency, $QE$, and short-circuit current density, $J_{SC}$, of the solar cell. For state-of-the-art devices we can assume that under reverse or short-circuit conditions, an ideal extraction of photo-generated charge carriers can be considered from absorber layers of the solar cell and the contributions from thin doped layers (such as p-a-Si:H and n-a-Si:H in amorphous silicon solar cell) can be neglected (Krause et al. 2001). In this case we can determine the external $QE$ as $QE(\lambda) = A_{absorber} \cdot (\lambda)$, where $A$ is the total absorptance of the absorber layer.

The second parameter that we use in simulations for characterization of external performance of the solar cells is the $J_{SC}$, which can be calculated as

$$J_{ph} = \sum_{i=1}^{i=K} \frac{q}{h \cdot c} \cdot S(\lambda_i) \cdot QE_{(opt)}(\lambda_i) \cdot \lambda_i \tag{3.33}$$

where $S(\lambda_i)$ is a discrete illumination spectrum (AM 1.5) and $q$ is an elementary charge ($q = 1.6 \cdot 10^{-19}$ As). The $QE$ can be replaced by $A_{abs}$, as discussed earlier.

If we are not satisfied with the results of optical and simplified electrical analysis, we must use a detailed electrical simulation tool, such as AMPS (Fonash et al. 1994), ASA (Zeman et al. 1997), ASPIN (Smole, Topič, and Furlan 1994), or Synopsys TCAD (Synopsys 2012).

## 3.4 DISCRETIZATION OF INPUT AND OUTPUT PARAMETERS IN NUMERICAL SIMULATIONS

In numerical simulations all input and output parameters are discretized into a certain mesh-grid. The parameters are presented with discrete values, representing the interval between the two neighbor discrete points. The mesh-grid should be dense enough to reproduce a realistic, usually continuous-like behavior of input and output parameters without noticeable deviations. We have to consider the expected dynamics of the input and output parameters. Whereas a discretization level that is too low can lead to noticeable deviations in simulated results, in comparison to realistic continuous functions, a discretization level that is too high results in time-consuming simulations. Practically, we can test an optimum in discretization level for a certain parameter in simulations by increasing the discretization level to the point where a saturation in the output results occurs.

In 1-D optical simulations an equidistant grid is often sufficient for spatial structure discretization. When performing 2-D or 3-D rigorous optical simulations, an adaptive grid is preferred in order to minimize the number of discrete points, affecting the simulation time.

Some recommended discretization levels for the example of simulation of thin-film silicon solar cells are listed here.

- Typical wavelength discretization: $\Delta\lambda = 5$ or $10$ nm
- Typical angular discretization: $\Delta\varphi = 2$ or $3$ degrees
- Typical spatial discretization:
  - Thin doped layers: $\Delta x = 1$ nm
  - Active (absorber) layers: 100 to 200 points per layer regardless of its thickness
  - Other supporting layers: 1 to 10 points per layer thickness

If we are interested only in absorptances in the entire layers, from which quantum efficiency can be calculated, and displaying the light intensity profile or the charge carrier generation profile inside a layer is not of interest, usually the discretization with one point at each interface is sufficient in 1-D optical simulations. Two related aspects must be considered: (1) In absorbing layers the light intensity can decrease a lot from one side to another side of the layer, and the exponential term in equations determining light propagation across the layer may be too large to be handled in the software we use. In this case, one solution is to introduce additional points inside the layer splitting the absorption into two or more parts. (2) We have to be aware that at one side of the interface the optical situation can be completely different from that on the other side. Therefore, it is wise to include two discrete points at the interfaces, practically being at the same position in the simulation structure, but one corresponding to the left side and the other to the right side of an interface.

## 3.5   RESULTS ON CALIBRATION AND VERIFICATION OF THE ONE-DIMENSIONAL SEMI-COHERENT OPTICAL APPROACH

In this section we present in detail two cases of calibration of scattering parameters and verification results for a 1-D semi-coherent optical modeling approach. We carried out the calibration of scattering parameters for two types of textured superstrates that were previously discussed: the Asahi U-type glass/$SnO_2$:F and glass/ZnO:Al superstrate. The superstrates have different types of surface texture and thus exhibit different scattering properties. Later on we present the results of simulations of amorphous silicon solar cells, considering the calibrated scattering parameters. We compare the results of simulations with the measurements of fabricated cells on the two types of superstrate. Verification results that we obtained for other thin-film solar cells are presented briefly at the end. In the next two sections the analysis is specifically devoted to the determination, comparison, and application of realistic scattering parameters as important input parameters in 1-D simulations of thin-film PV devices with textured interfaces.

### 3.5.1   DETERMINATION OF SCATTERING PARAMETERS AND SIMULATION OF AMORPHOUS SILICON SOLAR CELLS DEPOSITED ON ASAHI U-TYPE GLASS/$SnO_2$:F SUPERSTRATE

Asahi U-type $SnO_2$:F TCO superstrate is commonly used for amorphous silicon solar cells. The $SnO_2$:F TCO has a naturally grown surface texture. The thickness of the $SnO_2$:F layer in the analyzed sample was approximately 650 nm. The AFM scan and the SEM picture of the $SnO_2$:F textured surface are given in Figure 3.36. From AFM measurements we detected the vertical surface roughness $\sigma_{rms}$ of 40 nm. From both the AFM scan and the SEM picture we can observe pyramid-like texturization features.

By means of TIS measurements we determined the total and diffused reflectance and transmittance (Figure 3.37). In transmittance measurements

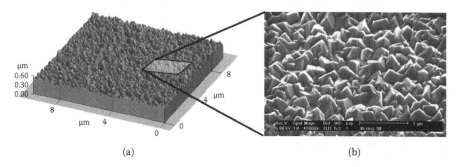

(a)                                                                (b)

**FIGURE 3.36**   (a) AFM scan and (b) SEM picture of the surface of the $SnO_2$:F layer of the Asahi U-type TCO superstrate.

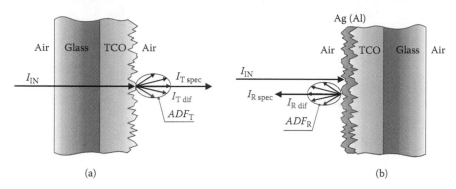

(a)                                                                    (b)

**FIGURE 3.37**   A schematic representation of the structure of a TCO superstrate. Optical situation in the determination of scattering parameters for (a) transmitted light and (b) reflected light.

we applied the light from the glass side. (Figure 3.37a). For reflectance measurement we covered the $SnO_2$:F surface with a thin layer of Ag ($d = 100$ nm) to increase the signal from the detector while keeping the texture almost unchanged (the texture of the $SnO_2$:F surface is transferred to the Ag surface). Higher reflectivity properties of the Ag surface enable more accurate determination of scattering properties. We illuminate the sample in this case from the Ag layer side (Figure 3.37b).

Whereas the $T_{tot}$ and $R_{tot}$ are mainly determined by the complex refractive indexes of the layers and surrounding material, the $T_{dif}$ and $R_{dif}$ depend on light scattering at the textured interface (Figure 3.38). We can observe a decrease of $T_{tot}$ at wavelengths below 450 nm in the measurements, which is a consequence of the higher absorption at shorter wavelengths in the $SnO_2$:F layer. The decrease in $R_{tot}$ at shorter wavelengths is a consequence of a lower reflectivity and higher absorption related to plasmonic effects of the Ag covering material in this wavelength region (especially the minimum peak at $\lambda = 325$ nm).

From $T_{tot}$, $T_{dif}$ and $R_{tot}$, $R_{dif}$ we determine the haze parameters $H_T$ and $H_R$, respectively. The results of $H_T$ are plotted in Figure 3.39. The $H_T$ values obtained from the measurements are plotted by circles, whereas the curves relate to calculations.

Decreasing the wavelength increases the haze, indicating that the scattering process at shorter wavelengths is more pronounced than at the long wavelength region. For short-wavelength light, the irregularities of texturization present a higher obstacle to trickle through specularly.

In Figure 3.39 we can observe deviations between the $H_T$ obtained with the measurement and the calculated one, if no correction is used in equations of scalar scattering theory (dashed curve in Figure 3.39). As we mentioned in Section 3.2.6.5, the deviations are assigned to the violations of the validity conditions of scattering theory (in particular the first one, $L \gg \lambda$). Therefore, we use the calibrated equations with the respective calibration function $c_T(\lambda)$ at $\sigma_{rms} = 40$ nm (dash-dot curve). We obtain the functions from the fitting

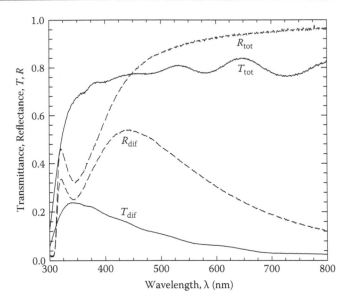

**FIGURE 3.38**  Wavelength-dependent transmittances and reflectances of the Asahi U-type SnO$_2$:F superstrate measured in air. In the case of reflectance measurement, the top surface of the SnO$_2$:F layer was covered with a thin Ag layer as shown in Figure 3.37.

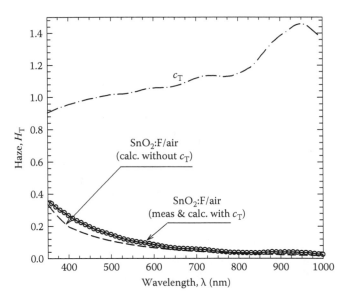

**FIGURE 3.39**  Haze parameter $H_T$ of the Asahi U-type SnO$_2$:F superstrate: $H_T$ obtained from the measurements (circles), calculated $H_T$ and the correction function $c_T$ (curves).

procedure of the calculated $H_T$ to the measured one. Thus, we can obtain a more accurate description of realistic $H_T$ (curve through the circles). The $c_T$ curve may be at longer wavelengths affected by uncertainties in measurements of small values of diffused signal in this wavelength region. However, changes in $c_T$ at longer wavelengths do not affect the $H_T$ values drastically in the absolute scale, since the $H_T$ is attenuated by the exponential function. In calculations of $H_T$, the power factor $a = 3$ was found to describe the slope of the exponential decay of $H_T$ relatively well.

In Figure 3.40 the haze parameter for reflected light $H_R$ obtained from $R_{tot}$ and $R_{dif}$ is plotted by circles. A comparison of $H_T$ and $H_R$ parameters for the same substrate reveals the higher values of $H_R$ (despite a small decrease of $\Delta\sigma_{rms} = 3$ nm due to 100-nm thick Ag covering). Higher $H_R$ values are not a consequence of high reflectance due to Ag covering since the haze is defined as a normalized parameter to $R_{tot}$ ($H_R = R_{dif}/R_{tot}$). Higher $H_R$ determined from the measurements indicates that the scattering process is more pronounced in reflection than in transmission.

The calculated $H_R$ with the original and calibrated equation and the respective calibration function $c_R(\lambda, \sigma_{rms})$ are plotted by curves in Figure 3.40. We can observe noticeable deviations between measured and calculated $H_R$, if no calibration function is used, especially in the long-wavelength region.

Considering the haze calibration functions obtained from the measurements on the TCO superstrate in air, one can use the equations of scalar scattering

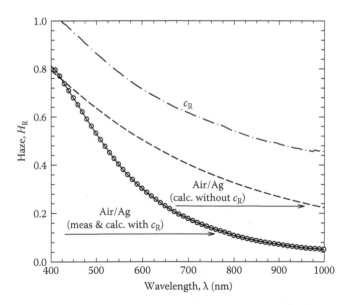

**FIGURE 3.40**   Haze parameter $H_R$ of the Asahi U-type SnO$_2$:F superstrate: $H_R$ obtained from the measurements (circles), calculated $H_R$ and the correction function $c_R$ (curves).

theory together with the calibration functions $c_T$ and $c_R$ to estimate the $H_T$ and $H_R$ at internal textured interfaces in the multilayer model. In the case of the analyzed amorphous silicon solar cell, the internal textured interfaces are $SnO_2$:F/p-a-SiC:H, p-a-SiC:H/i-a-Si:H, i-a-Si:H/n-a-Si:H, and n-a-Si:H/Ag. At each interface we consider the corresponding refractive indexes of adjacent layers. Regarding the decrease in the initial $\sigma_{rms}$ value of internal interfaces due to subsequent layer deposition on the TCO textured surface, we measured that the $\sigma_{rms}$ is decreased by 3 nm per 100 nm of subsequent layer thickness. This was also taken into account in calculations of $H_T$ and $H_R$. The results of haze calculations are presented in Figure 3.41 and Figure 3.42. The measured $H_T$ and $H_R$ for the $SnO_2$:F/air and air/Ag interface are also added to the graphs. The corresponding $\sigma_{rms}$ of the interfaces in the structure with $i$-layer thickness $d_i =$ 450 nm are indicated. We can observe that the calculated $H_R$ and $H_T$ values of internal interfaces are higher than the ones measured in surrounding air, which is the consequence of higher refractive indexes of layers compared to air.

Next, for optical simulations of the solar cell structure we determine the $ADF$s of the $SnO_2$:F Asahi U-type superstrate by the ARS system. The measured normalized $ADF_T$s versus scattering angle are shown in Figure 3.43.

The $ADF_T$s that correspond to two different angles of incident illumination $\varphi_{inc} = \varphi_{spec} = 0$ degrees and 60 degrees are shown. For $\varphi_{spec} = 0$ degrees, the measurements are shown for two different wavelengths of the laser beam: $\lambda = 633$ nm (open symbols) and $\lambda = 533$ nm (full symbols) as well. No noticeable

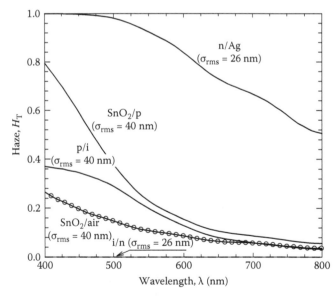

**FIGURE 3.41** The $H_T$ parameter determined for internal interfaces of the analyzed a-Si:H solar cell in a superstrate configuration. The measurement and the calculation for the $SnO_2$:F/air interface (superstrate characterization) are also added to the graph.

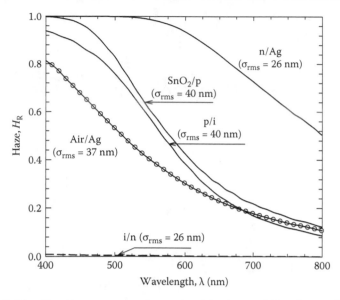

**FIGURE 3.42**   The $H_R$ parameter determined for internal interfaces of the analyzed a-Si:H solar cell in the superstrate configuration. The measurement and the calculation for the air/Ag interface are also added to the graph.

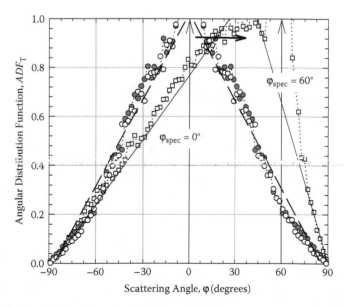

**FIGURE 3.43**   $ADF_T$ of the Asahi U-type $SnO_2$:F superstrate measured with red laser $\lambda = 633$ nm (open symbols) and green laser $\lambda = 532$ nm (full symbols) for two different angles of transmitted specular beam: $\varphi_{spec} = 0$ degrees and 60 degrees.

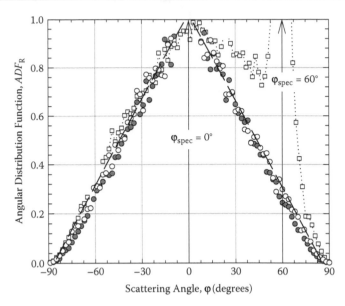

**FIGURE 3.44**   $ADF_R$ of the Asahi U-type SnO$_2$:F superstrate measured with red laser $\lambda$ = 633 nm (open symbols) and green laser $\lambda$ = 532 nm (full symbols) for two angles of reflected specular beam: $\varphi_{spec}$ = 0 degrees and $\varphi_{spec}$ = 60 degrees.

deviations between normalized $ADF_T$ corresponding to the two wavelengths are observed, indicating almost no wavelength dependency in this range. However, $ADF_T$ changes if $\varphi_{spec}$ is changed. The results show that by varying $\varphi_{spec}$, the peak of the $ADF_T$ shifts toward the corresponding specular direction as demonstrated in Figure 3.43 for $\varphi_{spec}$ = 60 degrees. Measurements indicated that only in the case of a normal incidence ($\varphi_{spec}$ = 0 degrees) is the peak of the $ADF_T$ located exactly at the specular direction.

To approximate the $ADF_T$ of the glass/SnO$_2$:F superstrates with an analytical description, a linear function presents as a good description.* For $\varphi_{spec} \neq 0$ degrees we can apply some higher order approximations if high accuracy is required. However, the simulations indicated no significant deviation in $QE$ if linear or more accurate higher order approximations of $ADF_T$ were used in this case.

The results of $ADF_R$ measurements of the substrate covered with Ag layer are given in Figure 3.44.

Also in this case no significant deviations between the $ADF_R$ obtained with red laser (open symbols) and green laser (full symbols) are observed in the figure for $\varphi_{spec}$ = 0 degrees. We can approximate the $ADF_R$ corresponding to the $\varphi_{spec}$ = 0 degrees with linear functions very well again. If $\varphi_{spec}$ is changed, the $ADF_R$ also changes but in contrast to the effect observed in

---

* This *ADF* is the function as measured in a plane; to include scattering in 3-D space we have to apply the transformation presented in Section 3.2.6.3.

$ADF_T$, in case of reflection the peak of the $ADF_R$ remains near to the scattering angle of 0 degrees regardless of the $\varphi_{spec}$. In Figure 3.44 this behavior is represented by the $ADF_R$ for $\varphi_{spec}$ = 60 degrees (open squares). We believe that the previously observed effects can be reproduced well by using advanced scattering models that enable us to calculate far-field scattering (Jäger and Zeman 2009).

The presented measured results we use further to calibrate equations and scattering parameters in optical simulations of solar cells. For the calibration of the haze parameter, we use the $c_T$ and $c_R$ functions as determined for measurements in air. $ADF$s were measured for four incident angles (0, 30, 60, and 75 degrees). For other angles between 0 degrees and 90 degrees we used linear interpolation. Such determined $ADF$s we applied to internal interfaces directly (no change as in the case of haze parameters). As discussed this might be one crucial approximation in the presented approach. Regarding complex refractive indexes we considered experimentally determined indexes of the layers (Zeman et al. 2000). Their real parts $n(\lambda)$ and imaginary ones $k(\lambda)$ are shown in Figures 3.45 and 3.46. In these simulations no reduction of the interface total reflectances was considered with respect to texturization. To simulate optical properties of a realistic n/Ag interface, we introduced a thin absorbing buffer layer ($d_{buff}$ = 1.5 nm) with optical properties of Al in the structures (Stiebig et al. 1994).

Single-junction a-Si:H solar cells in superstrate configuration glass/SnO$_2$:F (650 nm)/p-a-SiC:H (10 nm)/i-a-Si:H ($d_i$)/n-a-Si:H (20nm)/Ag were fabricated and characterized at Technical University Delft, the Netherlands. Samples

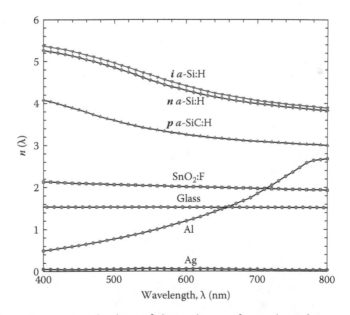

**FIGURE 3.45**  Measured values of the real part of complex refractive indexes, $n(\lambda)$, of the layers used in the simulations of the a-Si:H solar cells.

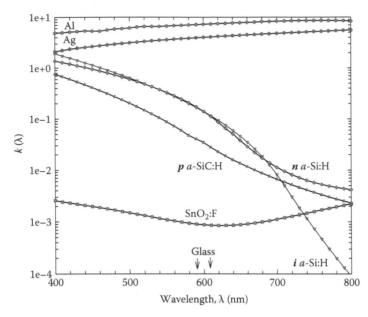

**FIGURE 3.46**  Measured values of the imaginary part of the complex refractive indexes, $k(\lambda)$, of the layers used in the simulations of the a-Si:H solar cells (note the logarithmic scale).

with four different intrinsic layer thicknesses $d_i$ = 150, 300, 450, and 600 nm were made.

We carried out optical simulations of the cells using the Sun*Shine* 1-D optical simulator. The simulated external quantum efficiency, *QE*, and short-circuit current density, $J_{SC}$, were calculated as described in Section 3.3. The results of simulated *QE* are compared to the *QE* obtained from the spectral response measurements of the cells in Figures 3.47 and 3.48. Solar cells were fabricated at TU Delft. We can observe a good agreement for all the solar cells with four different $d_i$. Measurements show and simulations reproduce very well the increase in *QE* especially at longer wavelengths if $d_i$ is increased. This is a consequence of higher absorption of light in a thicker *i*-layer. The intensity and position of the interference fringes observed in measured *QE* are matched with the simulated ones very well.

From the measured and calculated *QE* we determined $J_{ph}$ of the solar cells. We consider the illumination with the AM 1.5 spectrum in the wavelength interval from 400 to 800 nm. The $J_{SC}$ results are summarized in Table 3.1.

In increasing $d_i$, the $J_{SC}$ increases as a consequence of enhanced long-wavelength absorption in the intrinsic layer. Again we observe a good agreement between the measurements and simulations as a consequence of calibration of optical simulator to realistic scattering parameters and optical constants of the layers.

Another important result of optical simulations is charge carrier generation rate profile, $G_L$, which presents an input parameter for further detailed electrical simulations of solar cells. For electrical simulations of thin-film solar

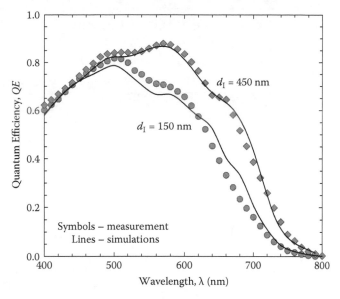

**FIGURE 3.47** Simulated (curves) and measured (symbols) quantum efficiency, *QE*, of the a-Si:H solar cells deposited on Asahi U-type SnO$_2$:F superstrate ($\sigma_{rms}$ = 40 nm) with two different thicknesses of the intrinsic layer, $d_i$.

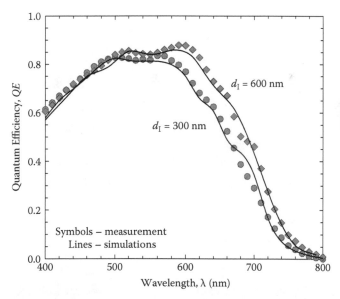

**FIGURE 3.48** Simulated (curves) and measured (symbols) quantum efficiency, *QE*, of the a-Si:H solar cells deposited on the Asahi U-type SnO$_2$:F superstrate ($\sigma_{rms}$ = 40 nm) with another two different thicknesses of the intrinsic layer, $d_i$.

**TABLE 3.1** **The Short-Circuit Current Density, $J_{SC}$, Determined from the Measured ($Q_{meas}$) and Simulated ($QE_{sim}$) Quantum Efficiencies of the a-Si:H Solar Cells Deposited on Asahi U-Type SnO$_2$:F Superstrate**

| | $J_{SC}$ (mA/cm²) under AM 1.5 Spectrum | |
| --- | --- | --- |
| $d_i$ (nm) | From $QE_{meas}$ | From $QE_{sim}$ |
| 150 | 12.38 | 12.68 |
| 300 | 14.37 | 14.40 |
| 450 | 15.57 | 15.51 |
| 600 | 15.88 | 15.72 |

cells, electrical simulators such as ASPIN, ASA, AMPS, or others can be used. In Figure 3.49, the calculated $G_L$s as a result of optical simulations are shown for selected wavelengths and for the entire AM 1.5 spectrum for the solar cell with $d_i$ = 450 nm. For shorter wavelengths (e.g., $\lambda$ = 350 nm), a steep decrease is observed in $G_L$ in the front part of the structure. This is because shorter

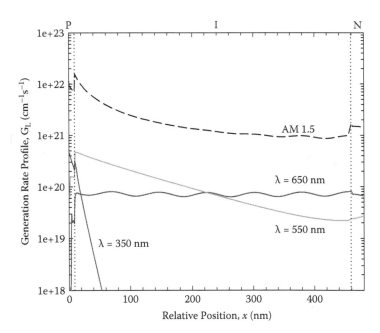

**FIGURE 3.49** Charge carrier generation rate profiles calculated in the p-i-n structure of the amorphous silicon solar cell with $d_i$ = 450 nm under 100 mW/cm² AM 1.5 light intensity and its spectral components (with 10-nm wavelength discretization step).

wavelengths experience high absorption due to large values of extinction coefficient of the materials in this wavelength region. Therefore, they are absorbed rapidly. For longer wavelengths (e.g., $\lambda = 650$ nm) the $G_L$ shows more equal distribution of charge generations throughout the structure. In this case we can observe small interference fringes in the $G_L$ as a consequence of interaction between forward- and backward-going waves of specular light. Specular light is only partially present in the structure, with textured interfaces. When summing up the $G_L$ contributions of all discrete wavelength intervals, we can calculate a common $G_L$ for the entire AM 1.5 illumination.

## 3.5.2   DETERMINATION OF SCATTERING PARAMETERS AND SIMULATION OF AMORPHOUS SILICON SOLAR CELLS DEPOSITED ON GLASS/ZnO:Al SUPERSTRATE

ZnO-based TCO superstrates present an alternative to glass/SnO$_2$:F superstrates for application in thin-film silicon photovoltaic devices. Two of the representatives are low-pressure chemical vapor deposited (LPCVD) boron doped zinc oxide, ZnO:B (Faye et al. 2007) and magnetron sputtered aluminum doped zinc oxide, ZnO:Al (Kluth et al. 1999). While the first one exhibits the natural surface texture as a consequence of poly-crystalline growth of the ZnO:B material in the LPCVD process, the texture of the ZnO:Al TCO is obtained by chemical wet etching (typically in 0.5% HCl solution) after sputtering of the layer (Kluth et al. 2003). In this section we optically characterize and carry out simulations for the ZnO:Al TCO.

Samples with ZnO:Al films on glass were fabricated at Research Centre Juelich, Institute of Photovoltaics, Germany, and etched to make the desired roughness of the surface. After sputtering, the native $\sigma_{rms}$ of the ZnO:Al films was almost flat (~4 nm). We applied wet chemical etching in diluted HCl (0.5%), using different etching times ($t = 1$–$30$ s) to obtain samples with the surface roughnesses $\sigma_{rms} = 25$–$120$ nm. The etching process can also reduce the final layer thickness that has to be considered in optical simulations. The SEM pictures of selected ZnO:Al surfaces with different $\sigma_{rms}$ are presented in Figure 3.50. One can observe different surface morphology than in the case of Asahi U-type SnO$_2$:F superstrate. The SEM pictures reveal hole- and crater-like surface morphology that we obtain by the controlled wet etching process. The morphology of the TCO surface and optical and electrical properties of the TCO layers were not optimized in this case.

We used TIS measurements to determine $T_{tot}$, $T_{dif}$ and $R_{tot}$, $R_{dif}$ of the glass/ZnO:Al superstrates. For the reflectance measurements we covered the ZnO:Al textured surface with a 100-nm thick Ag film, as in the case of the SnO$_2$:F superstrate. Haze parameters $H_T$ and $H_R$ were determined for the superstrates with different $\sigma_{rms}$. The $H_T$ and $H_R$ measurements of selected superstrates are shown in Figure 3.51 and Figure 3.52. The results show that the haze increases with the increasing $\sigma_{rms}$ and decreases with increasing $\lambda$ for both transmitted and

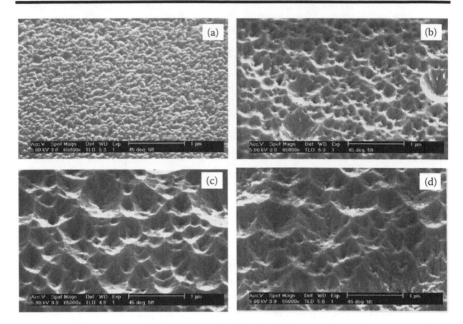

**FIGURE 3.50**    SEM pictures of the glass/ZnO:Al superstrates with (a) $\sigma_{rms} = 4$ nm, (b) $\sigma_{rms} = 35$ nm, (c) $\sigma_{rms} = 60$ nm, and (d) $\sigma_{rms} = 100$ nm. Different etching times were applied to achieve the roughnesses.

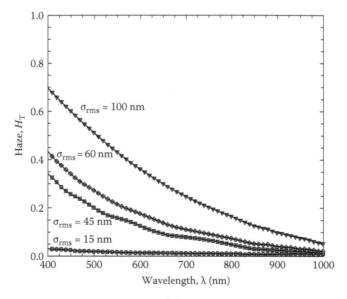

**FIGURE 3.51**    Haze parameters $H_T$ of the glass/ZnO:Al superstrate of different $\sigma_{rms}$: $H_T$ obtained from measurements (symbols) and calculations using calibration functions shown in Figure 3.53 (lines).

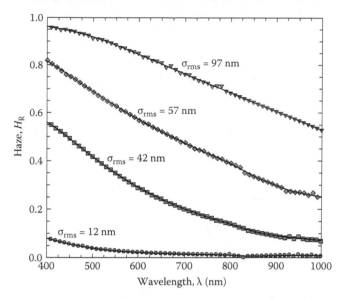

**FIGURE 3.52**   Haze parameters $H_R$ of the glass/ZnO:Al superstrate of different $\sigma_{rms}$: $H_R$ obtained from measurements (symbols) and calculations using calibration functions shown in Figure 3.54 (lines).

reflected light. A comparison of $H_T$ and $H_R$ parameters for the same superstrate reveals the higher values of $H_R$ again.

If we make a qualitative comparison between haze parameters of the Asahi U-type $SnO_2$:F (previous section) and ZnO:Al superstrates of similar $\sigma_{rms}$, we can observe deviations. In case of transmission we can assign some of the deviations to different complex refractive indexes of $SnO_2$:F and ZnO:Al layer. However, the deviations in $H_R$ of the interface formed by the same media (air/Ag) for both types of the superstrates indicate that in addition to the $\sigma_{rms}$ parameter and corresponding complex refractive indexes of the layers, other properties of interface morphology seem to affect the haze. Different morphologies of Asahi U-type (pyramids) and ZnO:Al superstrates (holes and craters) lead to deviations in haze parameters for the substrates with similar $\sigma_{rms}$.

We determined haze calibration functions in order to fit the measured $H_T$ and $H_R$ curves with the calculated ones (Figures 3.53 and 3.54). Since for superstrates with different $\sigma_{rms}$ we need different values of the calibration functions, a roughness dependency was introduced, resulting in $c_T(\lambda, \sigma_{rms})$ and $c_R(\lambda, \sigma_{rms})$. We can observe a decreasing trend for $c_T(\lambda, \sigma_{rms})$ by increasing $\sigma_{rms}$, whereas in the case of $c_R(\lambda, \sigma_{rms})$ no special trend can be observed with respect to the $\sigma_{rms}$. In the calculations of $H_T$ besides $c_T(\lambda, \sigma_{rms})$ the power factor $a = 3$ was used again.

Finding a way to determine the $c_T$ and $c_R$ analytically, without experiment, still presents a challenge.

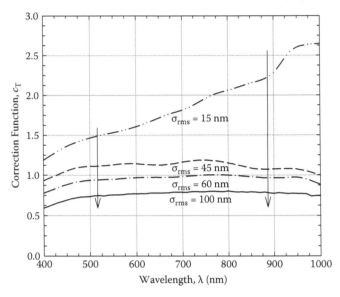

**FIGURE 3.53**    Correction functions $c_T$ for glass/ZnO:Al superstrates as a function of $\lambda$ and $\sigma_{rms}$.

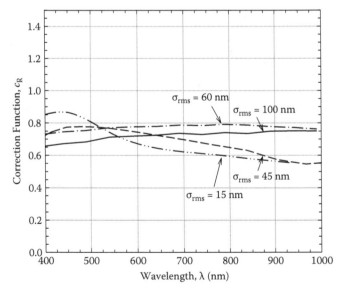

**FIGURE 3.54**    Correction functions $c_R$ for glass/ZnO:Al superstrates as a function of $\lambda$ and $\sigma_{rms}$.

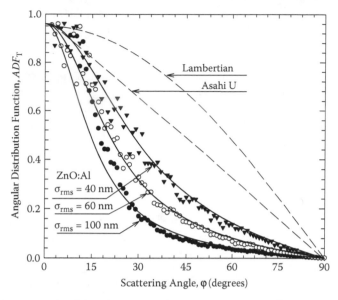

**FIGURE 3.55**    $ADF_T$ of the glass/ZnO:Al superstrates with different $\sigma_{rms}$: measurements (symbols), approximations (curves). The samples were illuminated with the red laser $\lambda = 633$ nm under normal incident angle.

The $ADF$s of three ZnO:Al samples for transmitted light, $ADF_T$, measured with the red laser ($\lambda = 633$ nm) are presented in Figure 3.55. Since we present the results for the perpendicular incidence of the laser beam, the $ADF_T$s are symmetrical around $\varphi = 0$. Therefore, only the results for positive scattering angles are presented in this graph. Besides the $ADF_T$ of the glass/ZnO:Al superstrates, the approximation of the $ADF_T$ of the Asahi U-type substrate (linear) and the $ADF_T$ of an ideal Lambertian diffuser are given for comparison. We can observe that increasing $\sigma_{rms}$, the normalized $ADF_T$ of the glass/ZnO:Al samples decreases or, from other point of view, that its shape narrows toward the specular direction ($\varphi = 0$ degrees). This means that for glass/ZnO:Al substrates with higher $\sigma_{rms}$, the scattering into smaller angles around the specular direction is relatively more pronounced. In substrates with lower $\sigma_r$, the light scatters more extensively into large angles ($\varphi > 30$ degrees). However, such representation of $ADF$ we can only use for relative comparison of light scattering with respect to the scattering angle, whereas the absolute intensity of scattered light is defined by the haze parameter.

A comparison of the $ADF_T$ of the glass/ZnO:Al superstrates with the $ADF_T$ of the Asahi U-type glass/SnO$_2$:F superstrate with $\sigma_{rms} = 40$ nm reveals that for the glass/ZnO:Al superstrates transmitted light is more extensively scattered into smaller angles around the specular direction than in the case of Asahi U-type superstrate. The Asahi U-type superstrate has a broader $ADF_T$. The $ADF_T$ of an ideal Lambertian diffuser, which is determined by cosine distribution function, is even broader than the $ADF_T$ of the Asahi U-type substrate. Therefore, the Lambertian distribution function, which is widely used in the optical modeling, is not good enough to approximate the realistic $ADF_T$ of the Asahi U-type and even less so of the ZnO:Al superstrates.

In Figure 3.56 the $ADF_T$s of the three glass/ZnO:Al superstrates are presented in a polar plot.

From this plot we can learn that the shape of the $ADF_T$s is very similar to the shape of a geometrical ellipse. Therefore, we can use the mathematical description of an ellipse as an analytical approximation of the $ADF_T$ of the glass/ZnO:Al superstrates (as measured in a 1-D plane). One of the possible mathematical descriptions of the ellipse is $r = r(\varphi, a, b)$ ($r$ – distance from the point where the ellipse crosses its horizontal axis at one side to a certain point on the ellipse; $\varphi$ – angle between horizontal axis and $r$; $a$ – horizontal radius; $b$ – vertical radius). Fixing the horizontal radius $a$ to unity and changing only one parameter, that is, vertical radius $b$, we can approximate the $ADF_T$s corresponding to different $\sigma_{rms}$ very well by the analytical function as shown in Figure 3.55 and Figure 3.56 by solid curves. To describe the $ADF_T$ corresponding to different $\sigma_{rms}$, $\varphi_{spec}$, and $\lambda$, we can use the following extended equation of an ellipse [Equation (3.34)]:

$$r = \frac{K \cdot 2b^2(\sigma_{rms}, \varphi_{spec}, \lambda)}{\cos\varphi(\sigma_{rms}, \varphi_{spec}, \lambda) \cdot \left[b^2(\sigma_{rms}, \varphi_{spec}, \lambda) + tg^2\varphi(\sigma_{rms}, \varphi_{spec}, \lambda)\right]} \tag{3.34}$$

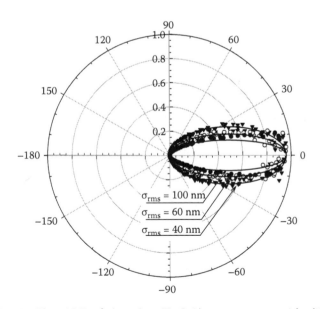

**FIGURE 3.56**   The $ADF_T$ of the glass/ZnO:Al superstrates with different $\sigma_{rms}$ represented in a polar plot.

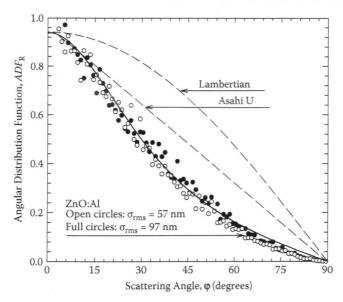

**FIGURE 3.57**   $ADF_R$ of the glass/ZnO:Al superstrates with different $\sigma_{rms}$: measurements (symbols), approximations (curves). The samples were illuminated with the red laser under normal incident angle.

The functions $b(\sigma_{rms}, \varphi_{spec}, \lambda)$ and $\varphi(\sigma_{rms}, \varphi_{spec}, \lambda)$ describe the dependency of the vertical radius $b$ and the angular parameter of the ellipse $\varphi$, respectively, on interface roughness $\sigma_{rms}$, angle of specular beam propagation $\varphi_{spec}$, and the wavelength $\lambda$. We determine all of these dependencies empirically. The factor $K$ is a normalizing constant. Using a polynomial function instead of the ellipse, we would need to introduce more parameters to obtain the required fit.

In Figure 3.57 we present the measurement results of the $ADF_R$ for the glass/ZnO:Al superstrates covered with the Ag layer, obtained with the red laser. In the case of $ADF_R$ no significant change in the distribution functions for superstrates with different $\sigma_{rms}$ was detected (demonstrated for two different $\sigma_{rms}$ in the figure). The ellipse was found again as a good analytical approximation of the $ADF_R$. From the comparison of the $ADF_R$ of the glass/ZnO:Al superstrates with the $ADF_R$ of Asahi U-type substrate and the Lambertian diffuser, which are also included in the figure, we can draw a conclusion similar to that of the $ADF_T$ analysis. At ZnO:Al superstrates, light scatters in narrower angles than in the Asahi U-type substrate or Lambertian diffuser. However, as observed from haze parameters, the scattering level can be much higher.

Further we show the results of investigation of a wavelength dependency of the $ADFs$ for the ZnO:Al samples (Figure 3.58). Again, only one half of the $ADF_T$ is shown (from 0 degrees to $\pm$ 90 degrees), but in this case the left and right sides of the figure correspond to two different $\sigma_{rms}$ of the superstrates. We can detect minor deviations between $ADF_T$ measured at $\lambda = 633$ nm and

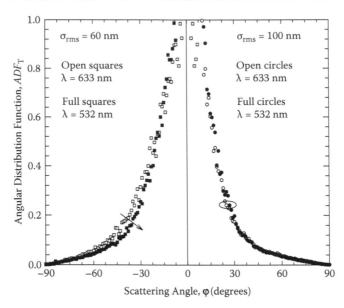

**FIGURE 3.58** $ADF_T$ of glass/ZnO:Al samples with two different $\sigma_{rms}$ illuminated by red ($\lambda = 633$ nm) and green ($\lambda = 532$ nm) layers.

at $\lambda = 532$ nm but only for the superstrates with $\sigma_{rms} \leq 60$ nm (left side of the figure). At higher $\sigma_{rms}$, no noticeable changes are detected (right side of the figure). In addition, measurements of the $ADF_R$ at $\lambda = 633$ nm and $\lambda = 532$ nm revealed no significant changes in the distribution functions (not shown here).

The measurements of $ADF_T$ and $ADF_R$ of the ZnO:Al samples under different incident angles of the laser beam indicated that in changing the incident angle, both the angle of outgoing (reflected and transmitted) specular beam ($\varphi_{spec}$) and the angular distribution of diffused light are changed. The results of $ADF_T$ and $ADF_R$ for $|\varphi_{spec}| = 60$ degrees and two different $\sigma_{rms}$ are plotted in Figure 3.59.

We observe the following effects from the results of the $ADF_T$:

■ The peak of the $ADF_T$ is not located at the specular direction as it is in the case of the perpendicular incidence.
■ There is no noticeable difference in the peak position of $ADF_T$ for the substrates with the two different $\sigma_{rms}$.
■ The $ADF_T$ corresponding to the larger $\sigma_{rms}$ is narrower than the $ADF_T$ of the smaller $\sigma_{rms}$ (indicated by the arrow), which is similar to the observation made in the case of the perpendicular illumination.

For the $ADF_R$ we can observe that the peak position of the $ADF_R$ is affected by the $\sigma_{rms}$ of the superstrate. With an increase in roughness, the deviation of

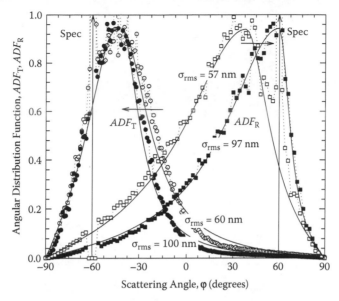

**FIGURE 3.59**    $ADF_T$ and $ADF_R$ of the glass/ZnO:Al superstrates with two different $\sigma_{rms}$, illuminated at $\varphi_{inc} = 60$ degrees.

the peak position from $\varphi_{spec} = 60$ degrees decreases (indicated by the arrow). The modified ellipses were found to be a good analytical approximation of the measured $ADF_T$ and $ADF_R$ for $|\varphi_{spec}| = 60$ degrees (solid lines in Figure 3.59) and were applied also for other non-perpendicular incident angles in simulations.

Calibration functions of haze parameters and $ADF$s were applied to internal textured interfaces of a-Si:H solar cells deposited on textured ZnO:Al superstrates. Again we used the Sun*Shine* optical simulator to carry out the simulations. For verification the solar cells were also fabricated on glass/ZnO:Al superstrates with different $\sigma_{rms}$ (4 [flat], 27, 45, and 100 nm) at Technical University Delft. The structure of the cells was glass/ZnO:Al ($d_{init} = 800$ nm)/ p-a-SiC:H (10 nm)/i-a-Si:H (450 nm)/n-a-Si:H (20nm)/Ag. The simulated and measured $QE$s of the cells are shown in Figure 3.60. By increasing the $\sigma_{rms}$ the following two effects are observed:

- The interference fringes are diminishing and are totally suppressed at a large interface roughness ($\sigma_{rms} = 100$ nm).
- The $QE$ at the longer wavelengths ($\lambda > 600$ nm) is enhanced.

These two effects are the consequences of an increased amount of scattered light and the decreased intensities of coherent, non-scattered light in the solar cells due to larger $\sigma_{rms}$ of the interfaces. We can clearly see these effects in Figure 3.60c, where the $QE$ of the solar cells with two different substrate $\sigma_{rms}$

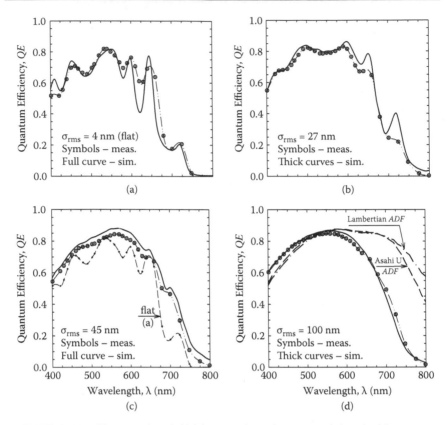

**FIGURE 3.60** The simulated (thick curves) and measured (symbols) quantum efficiency, *QE*, of the a-Si:H solar cells deposited on glass/ZnO:Al superstrates with (a) $\sigma_{rms} = 4$ nm – flat, (b) $\sigma_{rms} = 27$ nm, (c) $\sigma_{rms} = 45$ nm, and (d) $\sigma_{rms} = 100$ nm. Thin dash-dot-dot curves connecting the symbols are guidelines for the eyes.

(4 nm and 45 nm) are plotted in the same graph. In the graphs a good agreement between the measurements and simulations of the *QE* for all solar cells with different $\sigma_{rms}$ is demonstrated, indicating the validity of the model and the approach to determining the scattering parameters.

To demonstrate the importance of using the correct scattering parameters for the rough interfaces in a-Si:H solar cells deposited on the particular type of superstrate (in this case glass/ZnO:Al), we carried out additional simulations using other scattering parameters. If the correction functions $c_T(\lambda, \sigma_r)$ and $c_R(\lambda, \sigma_{rms})$ were not used, we observed the deviations in *QE* especially for small $\sigma_{rms}$ (below 40 nm), whereas for high $\sigma_{rms}$ the deviations diminish. However, the significance of using the correct *ADF* (*ADF*$_T$ and *ADF*$_R$) is demonstrated in Figure 3.60d. Using the *ADF*$_T$ and *ADF*$_R$ of the Asahi U-type substrate or generally used Lambertian distribution to simulate the *QE* of a-Si:H solar cells deposited on glass/ZnO:Al superstrates, we can observe a noticeable disagreement between simulations and measurements, especially at the longer wavelengths. Simulations indicate that in case of smaller $\sigma_{rms}$ the deviations are suppressed. These results demonstrate that for accurate optical simulations the realistic, well-calibrated scattering parameters are important.

Similar as for amorphous silicon solar cells we tested and verified the presented approach of modeling on other types of thin-film solar cells such as microcrystalline silicon, tandem micromorph, and chalcopyrite (Cu(In,Ga)Se$_2$) solar cells. Selected simulation results for these cells, where the calibration of scattering parameters was considered for the superstrates used (see specifications in figure captions), are presented in Figures 3.61, 3.62, and 3.63. More details on simulations of these solar cells can be found in Krč et al. (2006) and Zeman and Krč (2008).

To summarize, we highlighted the main input and output parameters of optical simulations of thin-film solar cells, focusing primarily on 1-D modeling. We mentioned basic techniques of determination of the input parameters. Special attention was paid to the determination of scattering parameters of nano-textured interfaces, which plays an important role in accurate 1-D simulations. We presented examples of simulations produced with the help of the Sun*Shine* optical simulator of a-Si:H solar cells deposited on two different superstrates. As indicated in Figures 3.61, 3.62 and 3.63 the presented modeling approach is well applicable also to other photovoltaic devices with nano-textured interfaces.

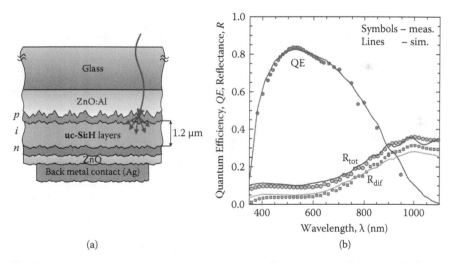

(a)    (b)

**FIGURE 3.61**    (a) A schematic representation of a microcrystalline (μc-Si:H) solar cell structure. (b) Simulated (lines) and measured (symbols) *QE*, $R_{tot}$ and $R_{dif}$ of the μc-Si:H solar cell with the i-μc-Si:H absorber layer thickness of 1.2 μm. The solar cell was deposited on glass/ZnO:Al textured superstrate with $\sigma_{rms}$ = 100 nm. The cell was fabricated at Research Center Juelich, Institute for Photovoltaics.

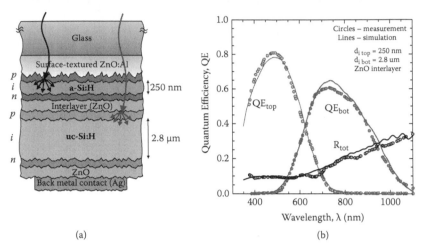

(a)                                          (b)

**FIGURE 3.62**    (a) A schematic representation of a micromorph (a-Si:H/µc-Si:H) double-junction solar cell structure. (b) Simulated (lines) and measured (symbols) *QE* and $R_{tot}$ of the a-Si:H/µc-Si:H solar cell with the thicknesses of absorber layers of 250 nm (i-a-Si:H) and 2.8 µm (i-µc-Si:H). The solar cell was deposited on glass/ZnO:Al textured superstrate with $\sigma_{rms}$ = 80 nm. The cell was fabricated at Research Center Juelich, Institute for Photovoltaics.

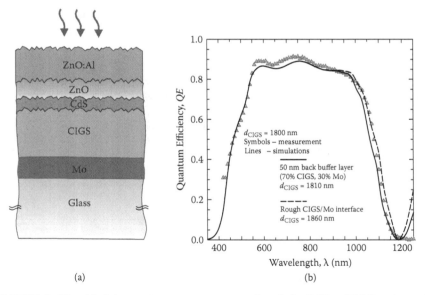

(a)                                          (b)

**FIGURE 3.63**    (a) Schematic representation of a Cu(In,Ga)Se$_2$ (CIGS) solar cell structure. The texture is introduced at the top interfaces by the polycrystalline CIGS absorber layer. (b) Simulated (lines) and measured (symbols) *QE* of the CIGS solar cell with the CIGS absorber layer thickness of 1.8 µm. The cell was fabricated at Angström Solar Center at Uppsala University.

# REFERENCES

Agashe, C., O. Kluth, G. Schöpe, H. Siekmann, J. Hüpkes, and B. Rech. 2003. "Optimization of the Electrical Properties of Magnetron Sputtered Aluminum-Doped Zinc Oxide Films for Opto-electronic Applications." *Thin Solid Films* 442 (1–2) (October 1): 167–172. doi:10.1016/S0040-6090(03)00966-0.

Beckmann, P., and A. Spizzichino. 1987. *The Scattering of Electromagnetic Waves from Rough Surfaces*. Artech Print on Demand.

Bennett, J. M., and L. Mattsson. 1999. *Introduction to Surface Roughness and Scattering*. 2nd ed. Optical Society of America.

Bie, Q. S., B. K. Cheong, M. K. Chung, Z. S. Lin, T. S. Lee, W. K. Kim, and S. G. Kim. 2000. "Determination of Optical Constants of Thin Films from Measurements of Reflectance and Transmittance." *Japanese Journal of Applied Physics Part 1— Regular Papers Short Notes & Review Papers* 39 (9A) (September): 5139–5143. doi:10.1143/JJAP.39.5139.

Bronshtein, I. N., K. A. Semendyayev, G. Musiol, and H. Mühlig. 2007. *Handbook of Mathematics*. 5th ed. Springer.

Carniglia, C. K. 1979. "Scalar Scattering Theory for Multilayer Optical Coatings." *Optical Engineering* 18: 104–115.

Choy, T. C. 1999. *Effective Medium Theory: Principles and Applications*. Oxford University Press.

Dominé, D., F.-J. Haug, C. Battaglia, and C. Ballif. 2010. "Modeling of Light Scattering from Micro- and Nanotextured Surfaces." *Journal of Applied Physics* 107 (4) (February 15). doi:10.1063/1.3295902.

Escarré, J., K. Söderström, C. Battaglia, F.-J. Haug, and C. Ballif. 2011. "High Fidelity Transfer of Nanometric Random Textures by UV Embossing for Thin Film Solar Cells Applications." *Solar Energy Materials and Solar Cells* 95 (3) (March): 881–886. doi:10.1016/j.solmat.2010.11.010.

Faye, S., J. Steinhauser, N. Oliveira, E. Vallat-Sauvain, and C. Ballif. 2007. "Opto-electronic Properties of Rough LP-CVD ZnO:B for Use as TCO in Thin-Film Silicon Solar Cells." *Thin Solid Films* 515 (24) (October 15): 8558–8561. doi:10.1016/j.tsf.2007.03.130.

Fonash, S., J. Hou, F. Rubinelli, M. Bennett, S. Wiedeman, L. Yang, and J. Newton. 1994. "Photodetection Enhancement in 2-Terminal Amorphous Silicon-Based Devices: An Experimental and Computer-Simulation Study." *Optical Engineering* 33 (6) (June): 2065–2069. doi:10.1117/12.169735.

Gracin, D., J. Sancho-Paramon, K. Juraič, A. Gajovič, and M. Ceh. 2009. "Analysis of Amorphous-nano-crystalline Multilayer Structures by Optical, Photo-deflection and Photo-current Spectroscopy." *Micron* 40 (1) (January): 56–60. doi:10.1016/j.micron.2008.03.011.

Hagemann, V., S. Reichel, S. Bauer, and P. Lechner. 2008. "Quantifying the Antireflection Properties of Internal Interfaces of Solar Cells on Rough TCO." Proc. of 23rd European Photovoltaic Solar Energy Conference and Exhibition, Valencia, Spain, September 1–5, WIP-Renewable Energies, pp. 2400–2402.

Holovský, J., A. Poruba, A. Purkrt, Z. Remeš, and M. Vaneček. 2008. "Comparison of Photocurrent Spectra Measured by FTPS and CPM for Amorphous Silicon Layers and Solar Cells." *Journal of Non-Crystalline Solids* 354 (19–25) (May 1): 2167–2170. doi:10.1016/j.jnoncrysol.2007.09.106.

IEC 60904-3:2008. "IEC 60904-3:2008—Photovoltaic Devices—Part 3: Measurement Principles for Terrestrial Photovoltaic (PV) Solar Devices with Reference Spectral Irradiance Data."

Isabella, O., J. Krč, and M. Zeman. 2010. "Modulated Surface Textures for Enhanced Light Trapping in Thin-Film Silicon Solar Cells." *Applied Physics Letters* 97 (10) (September 6). doi:10.1063/1.3488023.

Isabella, O., F. Moll, J. Krč, and M. Zeman. 2010. "Modulated Surface Textures Using Zinc-Oxide Films for Solar Cells Applications." *Physica Status Solidi a— Applications and Materials Science* 207 (3) (March): 642–646. doi:10.1002/pssa. 200982828.

Jäger, K., O. Isabella, R. A. C. M. M. van Swaaij, and M. Zeman. 2011. "Angular Resolved Scattering Measurements of Nano-textured Substrates in a Broad Wavelength Range." *Measurement Science & Technology* 22 (10) (October). doi:10.1088/0957-0233/22/10/105601.

Jäger, K., and M. Zeman. 2009. "A Scattering Model for Surface-Textured Thin Films." *Applied Physics Letters* 95 (17) (October 26). doi:10.1063/1.3254239.

Kelley, C. T. 1987. *Solving Nonlinear Equations with Newton's Method.* Society for Industrial Mathematics.

Kluth, O., B. Rech, L. Houben, S. Wieder, G. Schöpe, C. Beneking, H. Wagner, A. Löffl, and H. W. Schock. 1999. "Texture Etched ZnO:Al Coated Glass Substrates for Silicon Based Thin Film Solar Cells." *Thin Solid Films* 351 (1–2): 247–253. doi:10.1016/S0040-6090(99)00085-1.

Kluth, O., G. Schöpe, J. Hüpkes, C. Agashe, J. Müller, and B. Rech. 2003. "Modified Thornton Model for Magnetron Sputtered Zinc Oxide: Film Structure and Etching Behaviour." *Thin Solid Films* 442 (1–2) (October 1): 80–85. doi:10.1016/S0040-6090(03)00949-0.

Kong, J. A. 1990. *Electromagnetic Wave Theory.* 2nd ed. Wiley-Interscience.

Krause, M., M. Topič, H. Stiebig, and H. Wagner. 2001. "Thin-Film UV Detectors Based on Hydrogenated Amorphous Silicon and Its Alloys." *Physica Status Solidi a—Applications and Materials Science* 185 (1) (May 16): 121–127. doi:10.1002/1521-396X(200105)185:1<121::AID-PSSA121>3.0.CO;2-I.

Krč, J., G. Cernivec, A. Čampa, J. Malmstrom, M. Edoff, F. Smole, and M. Topič. 2006. "Optical and Electrical Modeling of Cu(In,Ga)Se-2 Solar Cells." *Optical and Quantum Electronics* 38 (12–14) (September): 1115–1123. doi:10.1007/s11082-006-9049-1.

Krč, J., M. Zeman, O. Kluth, E. Smole, and M. Topič. 2003. "Effect of Surface Roughness of ZnO: Al Films on Light Scattering in Hydrogenated Amorphous Silicon Solar Cells RID A-5194-2008." *Thin Solid Films* 426 (1–2) (February 24): 296–304. doi:10.1016/S0040-6090(03)00006-3.

Krč, J., M. Zeman, F. Smole, and M. Topič. 2002. "Optical Modeling of a-Si:H Solar Cells Deposited on Textured Glass/SnO2 Substrates RID A-5194-2008." *Journal of Applied Physics* 92 (2) (July 15): 749–755. doi:10.1063/1.1487910.

Macleod, H. A.. 2010. *Thin-Film Optical Filters.* 4th ed. CRC Press.

Ohlidal, I., K Navratl, and M. Ohlidal. 1995. "Scattering of Light from Multilayer System with Rough Boundaries." In *Progress in Optics*, Vol. 34, ed. E. Wolf, 29–54. North Holland.

Palik, E. D., ed. 1991. *Handbook of Optical Constants of Solids*, Vol. 2. Academic Press.

Peiponen, K.-E., R. Myllylä, and A. V. Priezzhev. 2010. *Optical Measurement Techniques: Innovations for Industry and the Life Sciences.* Springer.

PerkinElmer, Inc. 2012. http://www.perkinelmer.com/Catalog/Product/ID/L950.

Rockstuhl, C., S. Fahr, F. Lederer, F.-J. Haug, T. Söderström, S. Nicolay, M. Despeisse, and C. Ballif. 2011. "Light Absorption in Textured Thin Film Silicon Solar Cells: A Simple Scalar Scattering Approach Versus Rigorous Simulation." *Applied Physics Letters* 98 (5): 051102. doi:10.1063/1.3549175.

Rodriguez, J. M., R. S. Austin, and D. W. Bartlett. 2012. "A Method to Evaluate Profilometric Tooth Wear Measurements." *Dental Materials* 28 (3) (March): 245–251. doi:10.1016/j.dental.2011.10.002.

Sap, J. A, O. Isabella, K. Jäger, and M. Zeman. 2011. "Extraction of Optical Properties of Flat and Surface-Textured Transparent Conductive Oxide Films in a Broad Wavelength Range." *Thin Solid Films* 520 (3) (November 30): 1096–1101. doi:10.1016/j.tsf.2011.08.023.

Schmidt, J. A., R. R. Koropecki, R. D. Arce, F. A. Rubinelli, and R. H. Buitrago. 2000. "Energy-Resolved Photon Flux Dependence of the Steady State Photoconductivity in Hydrogenated Amorphous Silicon: Implications for the Constant Photocurrent Method." *Thin Solid Films* 376 (1–2) (November 1): 267–274. doi:10.1016/S0040-6090(00)01193-7.

Schropp, R. E. I., and M. Zeman. 1998. *Amorphous and Microcrystalline Silicon Solar Cells: Modeling, Materials and Device Technology.* Springer.

Smole, F., M. Topič, and J. Furlan. 1994. "Amorphous-Silicon Solar-Cell Computer-Model Incorporating the Effects of TCO/a-Si:C:H Junction." *Solar Energy Materials and Solar Cells* 34 (1–4) (September): 385–392. doi:10.1016/0927-0248(94)90064-7.

Springer, J., A. Poruba, and M. Vaneček. 2004. "Improved Three-Dimensional Optical Model for Thin-Film Silicon Solar Cells." *Journal of Applied Physics* 96 (9) (November 1): 5329–5337. doi:10.1063/1.1784555.

Stiebig, H., T. Brammer, T. Repmann, O. Kluth, N. Senoussaoui, A. Lambertz, and H. Wagner. 2000. "Light Scattering in Microcrystalline Silicon Thin Film Solar Cells." In *Proceedings of the Sixteenth European Photovoltaic Solar Energy Conference,* 549–552, James & James.

Stiebig, H., A. Kreisel, K. Winz, N. Schultz, C. Beneking, T. Eickhoff, H. Wagner, and M. Meer. 1994. "Spectral Response Modelling of a-Si:H Solar Cells Using Accurate Light Absorption Profiles." In *Photovoltaic Energy Conversion, 1994, Conference Record of the Twenty-Fourth IEEE Photovoltaic Specialists Conference, 1994,* 1:603–606. doi:10.1109/WCPEC.1994.520033.

Synopsys. 2012. "TCAD Simulation Tool." http://www.synopsys.com.

Tompkins, H. G. 2006. *A User's Guide to Ellipsometry.* Dover Publications.

Troparevsky, M. C., A. S. Sabau, A. R. Lupini, and Z. Zhang. 2010. "Transfer-Matrix Formalism for the Calculation of Optical Response in Multilayer Systems: From Coherent to Incoherent Interference RID B-9571-2008." *Optics Express* 18 (24) (November 22): 24715–24721. doi:10.1364/OE.18.024715.

Vaneček, M., and A. Poruba. 2002. "Fourier-Transform Photocurrent Spectroscopy of Microcrystalline Silicon for Solar Cells." *Applied Physics Letters* 80 (5) (February 4): 719–721. doi:10.1063/1.1446207.

Zeman, M., and J. Krč. 2008. "Optical and Electrical Modeling of Thin-Film Silicon Solar Cells." *Journal of Materials Research* 23 (4) (April): 889–898. doi:10.1557/JMR.2008.0125.

Zeman, M., R. A. C. M. M. van Swaaij, J. W. Metselaar, and R. E. I. Schropp. 2000. "Optical Modeling of a-Si:H Solar Cells with Rough Interfaces: Effect of Back Contact and Interface Roughness." *Journal of Applied Physics* 88 (11) (December 1): 6436–6443. doi:10.1063/1.1324690.

Zeman, M., J. A. Willemen, L. L. A. Vosteen, G. Tao, and J. W. Metselaar. 1997. "Computer Modelling of Current Matching in a-Si:H/a-Si:H Tandem Solar Cells on Textured TCO Substrates." *Solar Energy Materials and Solar Cells* 46 (2) (May): 81–99. doi:10.1016/S0927-0248(96)00094-3.

# Two- and Three- Dimensional Optical Modeling

4

## 4.1   INTRODUCTION

One-dimensional (1-D) optical modeling provides a powerful tool for analyzing optical aspects of thin-film photovoltaic devices. However, if different materials are introduced in the lateral direction, such as in device lateral structuring or surface texturing, certain assumptions and approximations have to be considered with respect to light scattering (scalar scattering theory) and antireflection. Another limitation of 1-D approaches is that the structure and applied illumination are assumed to be infinite in lateral directions; thus, effects at the edges of the device are not included in the analysis. If we want to consider variations in both lateral and vertical geometry of the structure, then two- (2-D) and three-dimensional (3-D) optical modeling become important.

In recent publications on optical modeling of thin-film photovoltaic devices, 2-D and 3-D modeling have been used primarily to

- resolve the situation of light scattering at either periodic (Haase and Stiebig 2007; Čampa et al. 2008; Dewan et al. 2011; Isabella et al. 2012a, 2012b) or random nano-textured interfaces (Rockstuhl et al. 2010; Bittkau et al. 2011; Jandl, Hertel, and Pflaum 2010, 2011),
- determine local near-field absorption inside individual layers of the structure (Bittkau and Beckers 2010), and
- investigate novel concepts such as solar cells with integrated metal nano-particles (Ferry, Polman, and Atwater 2011), nano-rods, and solar cells with high aspect ratios of surface textures (Chang, Hsieh, and Liu 2012).

Rigorous methods of solving electromagnetic wave equations inside the structures are applied, including the finite difference time domain (FDTD), finite element method (FEM), finite integration technique (FIT), and rigorous coupled-wave analysis (RCWA), all of which were mentioned in Chapter 1. Especially in 3-D simulations, the common bottleneck can still present long computation time and computer memory requirements if simulating thin-film structures considering realistic morphology of textured interfaces. For example, if $N$ discrete points (elements) are used to define the mesh-grid in 1-D simulation, in 2-D and 3-D simulation the number of points increases to $N^2$ and $N^3$, keeping the same discrete element size. This can result in millions or even billions of points to be calculated in the structure. Multi-core and super computers are preferred in such simulations, especially if iterative optimization processes of the device are performed.

Recently, combinations of the approaches as well as combinations of 1-D, 2-D, and 3-D modeling concepts have become of interest, in order to develop powerful and effective tools for optical simulation of different (thin-film) photovoltaic devices. One of them is 3-D simulator CROWM (Combine Ray-Optics Wave-Optics Model) applicable to optoelectronics devices with thin and thick layers or large textures (Lipovšek, Krč, and Topič 2011). It can be applied to thin-film semiconductor or organic solar cells as well as hetero-junction (HIT) cells. Other models also have been reported (Lockau 2012). An example of the combination of 1-D and 2-D or 3-D modeling, which was briefly mentioned in Chapter 2, is 1-D semi-coherent optical modeling that can be used to calculate the propagation of light throughout the structure combined with 2-D or 3-D modeling that can be used to determine light-scattering characteristics of nano-textured interfaces. Furthermore, combinations of 1-D and 3-D models in which light escapes at the edges of small-area thin-film cells into a thick glass superstrate can be taken into account (Springer, Poruba, and Vaneček 2004). Other advanced combinations are being developed.

In this chapter we first present and discuss some particular aspects of 2-D and 3-D optical modeling. Then we focus on 2-D optical modeling and present a 2-D optical model we developed based on the FEM approach.

## 4.2    ASPECTS OF TWO- AND THREE-DIMENSIONAL OPTICAL MODELING

If we employ a 2-D or 3-D optical model in the analysis of thin-film photovoltaic devices, in addition to variation in the structure and light intensity in a vertical direction (1-D model), variations in a lateral direction can be considered. As a variation of the structure in a lateral direction, we often think about surface texturing, lateral device structuring, or limited lateral dimensions of the device. In a 2-D approach, however, in the second lateral direction (considering Cartesian representation, as indicated in Figure 4.1a, $x$ is the vertical, $y$ is the first lateral, and $z$ is the second lateral direction), the structure is assumed to have infinite dimension in $z$-axis, with no changes in the geometry or illumination. Talking about textures, if the surface changes are present in only one dimension we call these textures 1-D or linear. If the texturization pattern is periodic, we often call it 1-D (linear) grating. Examples of structures that can be analyzed with a 2-D model are shown in Figure 4.1.

As concerns illumination of the structure, a plane wave is most commonly used in simulations. However, spatially limited sources (in one dimension in 2-D and in two dimensions in 3-D simulations), such as Gaussian source with limited distribution of light intensity, can be used as well. Usually, perpendicular illumination is assumed in 2-D and 3-D simulations; however, models should

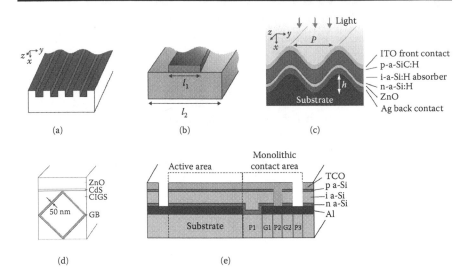

**FIGURE 4.1**   Examples of thin-film structures that can be analyzed with a 2-D modeling approach: (a) a structure with a 1-D periodic surface texture (grating) of rectangular shape—the surface variations are present in only one lateral direction (along the *y*-axis), (b) a structure with limited lateral dimensions, (c) amorphous silicon solar cell with sinusoidal 1-D periodic texture, (d) a chalcopyrite solar cell with a 2-D representation of grain boundary, (e) a thin-film silicon solar cell monolithically integrated in a PV module with lateral structuring. In all cases, infinite dimension of the structure is considered in the second lateral direction (*z*-axis). In the direction determined by the *y*-axis, the structure dimensions can be either finite (case b) or infinite (periodic, case a, c, d).

enable us to analyze also optical situations related to non-perpendicular illumination, which appears in outdoor conditions during the day. Furthermore, in most simulators based on rigorous solving of wave equations, the light is considered to be fully coherent, resulting in more intense interference fringes than in the case of measured spectral characteristics.

Let us proceed now to the aspects of 3-D modeling. Three-dimensional modeling needs to be used if the structure of a device or its surface texture or applied illumination changes in both lateral directions. Examples of structures in which 3-D modeling must be used are shown in Figure 4.2. However, as shown later, under certain assumptions 2-D modeling also can be employed quite effectively in some cases (to save on computation time and memory space) with tolerable uncertainty and predetermined and controlled loss of accuracy in the results.

Typical volumes (device segments) that are usually included in 3-D simulations are relatively small, in the range of a few cubic micrometers, and are

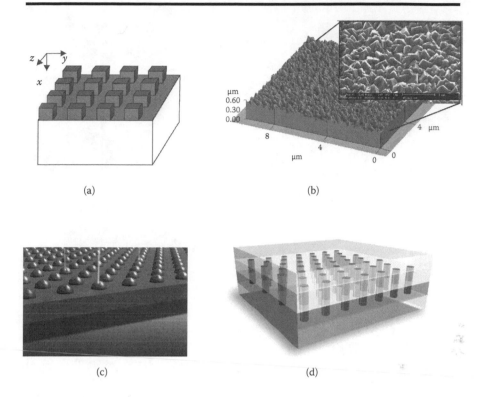

(a)  (b)

(c)  (d)

**FIGURE 4.2**  Examples of structures that need to be analyzed with 3-D models: (a) a structure with 2-D periodic surface texture (2-D grating)—the surface variations are present in both lateral directions (*y* and *z*), (b) a structure with 2-D random surface texture, (c) implementation of 3-D objects (metal nano-particles) in solar cell structure (example of a plasmonic solar cell, figure (Atwater and Polman 2010), (d) volume structuring of a photovoltaic device in both lateral directions (example of a nano-pillar solar cell (Kapadia et al. 2012).

limited by available computer processing power and memory space. Small volume is related to high-density discretization (mesh-grid) of the analyzed segment, which is required for accurate description of optical situation and geometrical description of nano-textures and nano-structures present in thin-film photovoltaic devices. Having the dimensions of textures or lateral structures in the range of 10 µm, one can use a geometrical optics approach rather than time-consuming rigorous solvers. In structures where certain periodicity in the range of micrometers can be recognized in lateral directions, limitations in the size of the segment that is included in the simulation is not critical, since only one or even a half period is enough to be included in the calculation if proper boundary conditions are set (see next section). For random nano-structures or textures that are introduced in the thin-film

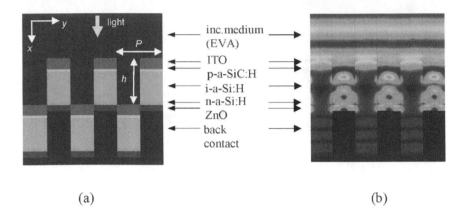

(a)                                           (b)

**FIGURE 4.3** (a) Cross-section of the analyzed 3-D structure of a-Si:H solar cell with 2-D periodic textures (as presented in Figure 4.2a); (b) results of 3-D optical simulations for $P = 300$ nm and $h = 300$ nm: absolute value of electric field ($\lambda = 600$ nm) in different layers of the a-Si:H cell.

devices, one would basically need to consider larger segments to preserve the statistical properties of the random character. This can quickly require enormous computer processing power and memory space; therefore, most 3-D simulations that are reported for random structures are limited to a few correlation lengths of random variations. It is also important how we select the segments from a random surface, representing the entire texture (Jandl et al. 2010).

Next, we show and discuss the results of two examples of 3-D simulations: first, a solar cell structure with periodic texture and, second, a structure with a random type of texturization. For the example with periodic surface texture, we present the calculated absolute value of the electric field of electromagnetic waves (at wavelength of $\lambda = 600$ nm) in an a-Si:H solar cell structure with periodically textured interfaces. In simulations we use the simulator COMSOL, which is based on the FEM simulation approach. The cross-section of the analyzed solar cell structure is schematically shown in Figure 4.3a. The cell has the following configuration: ITO front contact (70 nm)/p-a-SiC:H (15 nm)/i-a-Si:H absorber (200 nm)/n-a-Si:H (20 nm)/ZnO (40 nm)/ideal back contact. The texturization has 2-D periodic character, the shape of the features is rectangular (see Figure 4.2), and the period and height of the texturization features are $P = 300$ nm and $h = 300$ nm. The same texturization is applied to all interfaces in the solar cell structure. In this simulation we assume ideally steep vertical walls of the texturization, resulting in discontinuity of some layers in the simulated structure. In realistic devices the walls have certain inclination and thin coverage with layers (considered in simulations in the next chapter). In Figure 4.3b, the absolute value of the electric field is shown for a cross-section of the device ($x,y$ plane), as presented in Figure 4.3a. The regions with high and low intensity of the

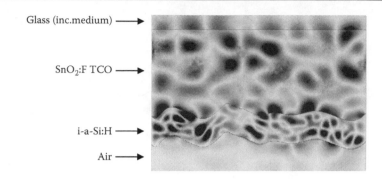

Glass (inc.medium) ⟶

SnO$_2$:F TCO ⟶

i-a-Si:H ⟶

Air ⟶

**FIGURE 4.4**   Calculated light intensity distribution ($\lambda = 600$ nm) in a cross-section of a structure ($x, y$ plane) with realistic random surface texture of SnO$_2$:F Asahi U transparent conductive oxide on glass.

field can be recognized. This picture can serve as a possible visual representation of the optical situation in the structure at a certain cross-section and certain wavelength of light; however, to determine the performance of the solar cell, the field at each point of the structure has to be determined for each wavelength of the illumination spectrum.

The second example of 3-D optical simulation is the structure with 2-D random texture. In Figure 4.4 the intensity of light, which was calculated based on the electric and magnetic field, using the Poynting vector, is shown for the structure glass (in simulations as incident medium)/SnO$_2$:F TCO (600 nm)/ i-a-Si:H layer (200 nm). Simulations were performed, again, with the COMSOL simulator. Bright areas in the picture indicate a high intensity of light.

For verification of such optical simulations, near-field scanning optical microscopy (NSOM) can be used in which the near-field intensity of light can be measured (Bittkau and Beckers 2010). However, the measurements can be done in air above the surface of the device. Therefore, the intensity distribution inside the structure still remains in the domain of the simulation only.

Because of the complexity and computer processing power related to 3-D analysis, we often try to describe our problem with a model that has fewer dimensions. As mentioned, 2-D or even 1-D modeling can be used much more effectively than 3-D modeling in the analysis and optimization, if the problem can be transferred to 2-D or 3-D. We show an example where 2-D optical modeling was used to detect the range of optimal periods and heights of the periodic surface texture in a-Si:H solar cells that was presented in Figure 4.3a. Basically the applied texture is supposed to be 2-D (Figure 4.2a), but with a 2-D model we can consider only a 1-D texture (Figure 4.1a). Simulations were carried out in this case with a 3-D MEEP FDTD simulator (MEEP simulator), considering the 2-D texture, and with a 2-D simulator FEMOS (Čampa, Krč, and Topič 2008) (see Section 4.3 for more detail on its development) considering a 1-D texture. Both types of simulations were carried out for the same cell

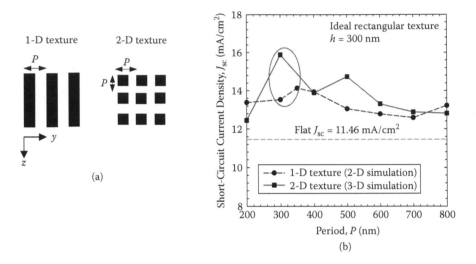

**FIGURE 4.5**    (a) Top view of 1-D and 2-D texture (left) and (b) the graph of calculated short-circuit current as a function of *P* for the cell with 1-D and 2-D texture (right).

configuration. Potential improvements in short-circuit currents with respect to the cell with flat interfaces were calculated for both cases of simulations, considering the absorptance in i-a-Si:H absorber layer, by applying an AM 1.5 solar spectrum (see Section 4.3 for a detailed explanation on short-circuit calculation from the absorptance in the i-a-Si:H layer). The results of $J_{SC}$ as a function of the period $P$ of the interface texture are shown in Figure 4.5. Although the results of the two simulations differ, we can observe that the same region of optimal periods (denoted with a circle), where we get the highest gain in $J_{SC}$, is obtained for both 2-D and 3-D simulations. The same region of optimal heights ($h \approx 300$ nm, not shown here) of the texturization feature was also identified for 2-D and 3-D simulations. In the simulation results that correspond to 3-D simulation of the structure with 2-D texture in Figure 4.5a, we can observe a distinct second maximum in $J_{SC}$ at $P \approx 500$ nm. The curve that corresponds to 2-D simulations of 1-D texture has the second maximum as well, but at longer wavelengths $\lambda >$ 800 nm. For additional validation of the results corresponding to 1-D texture, we also apply the 3-D model MEEP to the structure with 1-D texture. The same results were obtained as with the 2-D simulator FEMOS. Next we focus on the first maxima of the structures with 1-D and 2-D texture, that both appear in the narrow interval of periods $P = 300$–350 nm. Despite the same position, much larger improvements in $J_{SC}$ are obtained for 2-D textures than for 1-D, as revealed in Figure 4.5. However, by means of 2-D simulation we can determine the interval of expected optimal values, and then 3-D simulation still needs to be used for the selected periods to detect the maximal $J_{SC}$ gains corresponding to the structures with 2-D textures. The number of simulations performed by 3-D simulator to find the optimum is minimized in this way.

## 4.3  TWO-DIMENSIONAL OPTICAL MODEL BASED ON THE FINITE ELEMENT METHOD

In this section we present detailed aspects of development of a 2-D optical model that can be used for analysis of thin-film photovoltaic devices (Čampa 2009). The model uses the FEM to solve the wave equations in order to determine the optical situation in the structure. The model was developed primarily for thin-film silicon solar cell structures with periodically textured interfaces, but it is applicable to other types of solar cells and textures. An example of a typical solar cell structure that can be analyzed with the presented 2-D optical model is illustrated in Figure 4.1c. With this figure we also introduce the orientation of the Cartesian system in the model:

- the $x$-axis presents the vertical direction of the structure, as in the 1-D model presented in Chapter 2;
- the $y$-axis is the first lateral axis, in which the texture, structure, or incident illumination can be changed; and
- the $z$-axis is the second lateral axis, where no changes in the structure or illumination are assumed.

From among the methods of rigorous solving of wave equations, we selected the FEM for the following reasons:

- Basic discrete elements can be triangles (not squares), which facilitates an accurate approximation of realistic surface curvatures in thin-film solar cells. In addition, higher order approximation functions of the field inside the triangular element can be applied to accurately define the field value inside the discrete element (see Section 4.3.2).
- The method is used directly in the frequency (wavelength) domain, which is convenient since most of the input parameters and output results of simulations are defined as functions of light wavelength.
- Description of a complex optical system (multilayer structures with textured interfaces) is made simpler by using FEM instead of methods based on integral techniques, for example, where a boundary value problem must be resolved.
- Results of the near field determined inside a small simulation domain can be used to determine the far field outside the simulation domain, by using the Huygens principle (Jin 2002) or other methods, since the field at the boundaries of the simulation domain is well defined.
- FEM turned out to be a very efficient approach in simulations of thin-film structures where optical properties of realistic metal films or nano-structures are included. For example, plasmonic losses at textured Ag back reflector or plasmonic scattering at metal nano-particles, introduced in solar cells to boost the absorption in the active layers of solar cells, can be analyzed very efficiently by FEM.

## 4.3.1   ELECTROMAGNETIC BACKGROUND OF THE MODEL

We consider light as electromagnetic waves (fully coherent analysis), described by electric, $E$, and magnetic, $H$, field strengths.[*] Basically, $E$ and $H$ are time-harmonic vectors, defining the intensity of light (by the Poynting vector; see Chapter 2), the direction of propagation (defined by the direction of the real part of the Poynting vector), and the wavelength of light (by the frequency of harmonic oscillation and the velocity of light in the material). We usually apply the Fourier transform to $E$ and $H$ to obtain frequency-dependent complex vectors. By following Maxwell's equations in the frequency domain, we can derive vector wave equations [Equation (4.1)] for $E$ and $H$ (Kong 1990).

$$\nabla \times \left( \frac{1}{\mu} \nabla \times E \right) - \omega^2 \varepsilon E = -j\omega J$$

$$\nabla \times \left( \frac{1}{\varepsilon} \nabla \times H \right) - \omega^2 \mu H = \nabla \times \left( \frac{1}{\varepsilon} J \right) \tag{4.1}$$

Symbol $J$ presents the electric current density induced by electromagnetic oscillations. Symbols $\varepsilon$ and $\mu$ are complex permittivity (dielectric function) and permeability of the medium, whereas $\omega$ presents the angular frequency of light. The $\varepsilon$ and $\mu$ generally can be complex tensors; however, materials that we consider here have isotropic properties in all directions, and thus $\varepsilon$ and $\mu$ are presented by complex values.

Equations (4.1) are written in their general form; later, we apply them to a 2-D case where we assume that the fields are not changed along the $z$-axis. Additionally, we assume that $J$, which presents the induced electrical current in the range of electromagnetic oscillations, can be set to zero in the photovoltaic materials used. If we take these assumptions into account, we obtain scalar wave equations for transverse electric (TE) and transverse magnetic (TM) waves as given in Equation (4.2):

$$\left[ \frac{\partial}{\partial x} \left( \frac{1}{\mu_r} \frac{\partial}{\partial x} \right) + \frac{\partial}{\partial y} \left( \frac{1}{\mu_r} \frac{\partial}{\partial y} \right) + k_0^2 \varepsilon_r \right] E_z = 0$$

$$\left[ \frac{\partial}{\partial x} \left( \frac{1}{\varepsilon_r} \frac{\partial}{\partial x} \right) + \frac{\partial}{\partial y} \left( \frac{1}{\varepsilon_r} \frac{\partial}{\partial y} \right) + k_0^2 \mu_r \right] H_z = 0 \tag{4.2}$$

where $\varepsilon_r$ and $\mu_r$ are complex values representing the relative permittivity and relative permeability of the material and are connected with $\varepsilon$ and $\mu$ as $\varepsilon = \varepsilon_r \cdot \varepsilon_0$ ($\varepsilon_0 = 8.854 \cdot 10^{-12}$ F/m) and $\mu = \mu_r \cdot \mu_0$ ($\mu_0 = 4\pi \cdot 10^{-7}$ H/m). The $k_0$ is a wavenumber in free space and is defined as $k_0 = \omega \sqrt{\varepsilon_0 \mu_0}$.

---

[*] We use bold font to denote vectors.

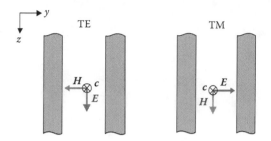

**FIGURE 4.6**   Definition of TE and TM polarization—orientation of E and H field with respect to the 1-D texture represented by two grating stripes.

As mentioned, the waves are divided into TE and TM polarized waves. Despite the perpendicular incident illumination (see Section 4.3.4), the waves have to be divided in the two groups because of the variations in lateral directions of the structure. How we define the TE and TM polarization with respect to 1-D grating is presented in Figure 4.6.*

In the case of TE polarization we can entirely represent the wave by the $E$ component in the $z$-direction ($E_z$), whereas for TM polarization the wave is fully represented by the $H$ component in the $z$-direction ($H_z$). Because the incident light usually includes both TE and TM polarization components, we have to consider both equations in Equation (4.2) in our calculations.

In simulations of thin-film solar cells we assume that the layers do not posess\ any magnetic characteristics, which leads to $\mu_r = 1$. Thus, the optical properties are entirely defined by complex permittivity $\varepsilon$, which is in most cases the wavelength-dependent function. Most commonly we use, instead of dielectric functions, the complex refractive index $N(\lambda) = n(\lambda) - jk(\lambda)$, as mentioned in Chapter 3.

## 4.3.2   DISCRETIZATION OF THE STRUCTURE

In numerical calculations we have to introduce a discretization of the structure (spatial discretization), of the illumination spectrum (wavelength discretization), and of other input and output quantities. In the presented 2-D model, which is based on FEM method, a triangular spatial discretization is used (Figure 4.7). The spatial region of the device (area in 2-D) that we include in simulation we assign to the simulation domain. The simulation domain is divided into basic discretization elements, according to the defined mesh-grid.

We often assume that the optical properties of the material inside the basic discretization element do not change or that they change in accordance with a

---

* We assume the incident plane to be parallel with the $xy$ plane, and the "T" (transverse) means transverse (rectangular) to this plane.

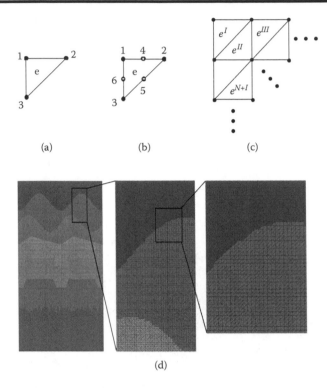

**FIGURE 4.7**    Basic triangular elements used in the 2-D FEM model: (a) element with nodes (1–3) corresponding to linear approximation of the field along the element, (b) element with nodes (1–6) corresponding to quadratic approximation of the field, (c) combination of basic elements with their nodes, and (d) examples of different interface morphology shapes and enlargements (center and right) for sinusoidal shape.

known function (linear, quadratic). To simplify the description of the structure and save on calculation time we use primarily an equidistant mesh-grid in our simulator, which means regular alignment of elements of the same size over the entire simulation domain (Figures 4.7c and 4.7d). In particular, the equidistant grid can simplify the matrixes in which the calculation data are stored, significantly reducing the calculation time. Unfortunately, if layers with the thickness of 10 μm up to a few millimeters (usually substrates), in combination with a thin-film multilayer structure, are present in the simulated structure the equidistant grid presents a drawback. Therefore, in solar cells in a superstrate configuration (glass at the front side), only a part of thick superstrate can be included in the calculation domain; otherwise, we have to use some other tricks.

Inside each triangular element, the field strength $E_z$ or $H_z$ is not assumed to be a constant value but may change, as mentioned. The values at the corners of the triangle are calculated when solving the system of equations describing the optical situation in the structure (see Section 4.3.5). Then we use either a linear or quadratic approximation to determine the field at any point inside the triangle. The mathematical representation of the linear and quadratic approximations is given in Equations (4.3) and (4.4), respectively. In these equations and in further descriptions the symbol $\phi$ stands for either $E_Z$ or $H_Z$, depending on the polarization analyzed. The superscript "e" denotes that a given quantity value corresponds to an element $e$. Coefficients $a$ through $f$ are determined based on the $\phi$ values at so-called nodes located at the borders of the triangle. In the case of linear approximation these nodes are located at the corner of the triangle. In the case of quadratic approximation we have to generate additional nodes in the middle of the triangular borders (Figure 4.7b). Each node corresponds to an unknown, which is included in the system of equations.

$$\phi^e(x,y) = a^e + b^e x + c^e y \tag{4.3}$$

$$\phi^e(x,y) = a^e + b^e x + c^e y + d^e x^2 + e^e xy + f^e y^2 \tag{4.4}$$

In practical cases of simulations of thin-film solar cells with textured interfaces, linear approximation is often sufficient.

Practically, in simulations of thin-film solar cells it is not the curvatures of the textured interfaces but rather the requirement of using a sufficient number of nodes per wavelength in the material that determines the size of the discrete elements (density of the mesh-grid). According to our experiences in simulations of thin-film solar cells, the element size (the short vertices of the triangle) should be less than 10 nm. For accurate study of optical effects around textured metals with realistic optical properties, this size should be less than 5 nm. Of course, the size also depends on the optical density (refractive index) of materials, determining the wavelength of the material.

### 4.3.3   BOUNDARY CONDITIONS

When we define the discretization grid and optical properties of layers in the simulation domain, we also have to apply proper boundary conditions at the left, right, top, and bottom boundaries of the 2-D domain (see Figure 4.8). With boundary conditions we set the rules for determination of the electric and magnetic field at the boundaries and implicitly also inside and outside of the simulation domain.

First we address the conditions at the left and right boundaries. These conditions are defined based on the lateral geometry of the structure and incident illumination inside and outside the simulation domain. Let us consider first the structures in which lateral geometry has a periodic nature and the applied illumination is a plane wave. Incident illumination will be applied under a perpendicular angle in all analyzed cases.

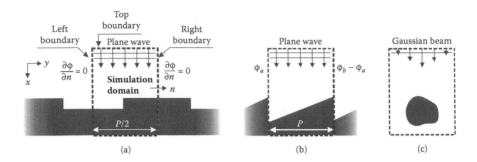

**FIGURE 4.8**  Simulation domains with different conditions at left and right boundary: (a) symmetrical periodic boundary condition, (b) general periodic boundary condition, and (c) absorbing boundary conditions (open region system).

Considering lateral periodicity, two types of *periodic boundary conditions* can be applied at the left and right boundaries of the simulation domain. The first type is used if the texture profile is symmetrical over the left and right boundaries (Figure 4.8a). In this case, the field $\phi$ ($E_z$ or $H_z$) has to be mirrored over the boundaries. Mathematically this means that the first derivative of $\phi$ to the normal direction of the boundary has to be set to zero. This boundary condition is known as a *symmetrical periodic boundary condition* or homogeneous Neumann condition (Jin 2002). It allows that in a lateral direction, a part of the structure corresponding to only a half of the period can be included in the simulation domain. This can save us a lot of computation time compared to the structures where the whole period has to be taken into account.

However, in many cases the symmetry of the texture features is not present. In this case, we should consider *general periodic boundary conditions* known as Dirichlet conditions (Figure 4.8b). Here, left and right boundaries are virtually connected, which means that the $\phi$ at the left and right sides of the boundaries have to be equal.

The third example of boundary condition is the *absorbing boundary condition,* which implies a so-called open region system (Figure 4.8c). In this case, the entire structure is included in the simulation domain. Thus, no periodicity of surface texture in lateral dimension is required. Here, the boundary condition has to ensure that light that approaches the boundary is efficiently absorbed there, ensuring negligible reflectivity (therefore, the name *absorbing boundary condition*). This imitates the escape of light out of the simulation domain (open system). However, it turns out we cannot achieve an ideal absorbing condition, that is, a boundary with zero reflectivity, but to approach this we can implement absorbing boundary conditions of higher orders (Jin 2002). To illustrate the situation, we present the reflectivity of first and second order absorbing

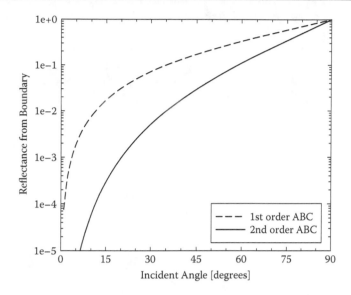

**FIGURE 4.9**   Reflectance of the wave at the boundaries, considering different absorbing conditions as a function of incident angle.

boundary conditions (Figure 4.9) as a function of the incident angle of the light that approaches the boundary. We can see that the second order condition has lower reflectivity for all incident angles in comparison to the first order condition. Furthermore, it approaches zero for the perpendicular incidence and gets close to unity at the oblique incident angle. According to our experience the use of the second order absorbing boundary condition leads to sufficiently small errors, related to this, in simulations of thin-film solar cell structures. However, we have to be careful that the structure where the light is scattered is not too close to the boundary (typically the structure should be away at least one wavelength). It is recommended to place the structure in the middle of the simulation domain and to take enough free space around to minimize the maximal possible incident angle on the boundaries to about 45 degrees. At this angle around 3% of the light is reflected back into simulation domain. In simulation of realistic cases, many light waves approach the boundary under angles that are usually below 45 degrees; thus, the overall error due to this reflection is much smaller than 3%.

So far we have addressed the three types of boundary conditions that can be applied to the left and right boundaries. What is the situation at the top and bottom boundaries? In this case, the structure is usually limited with the finite thicknesses; thus, almost always the absorbing boundary conditions, like the one described, are applied to the top and bottom boundaries.

### 4.3.4    INCIDENT ILLUMINATION

Two basic types of incident illumination that are considered in the presented optical model are plane wave and linear* source with Gaussian intensity distribution. In periodic boundary conditions we use the plane wave, whereas the linear Gaussian source is applied to an open system configuration. In simulation the plane wave can be used also as idealization of the waves that are planar only in a limited area. For instance, illumination from the sun or illumination from a solar simulator used in indoor testing can be approximated with a plane wave (Born and Wolf 1999; Lockau 2012).

The illumination sources in the presented model are generated at the top boundary together with the absorbing boundary condition. For example, if applying a plane wave with TE polarization, the equation considering absorbing condition in combination with the illumination is as follows:

$$-\frac{\partial E_z}{\partial x} + jk_0 E_z + \frac{j}{2k_0}\frac{\partial^2 E_z}{\partial y^2} = 2jk_0 E_{0z} \tag{4.5}$$

where $E_{0z}$ presents the electric field of the applied wave at the top boundary ($x = 0$). In this case only electric fields in the $z$-axis, corresponding to TE assignment, are present. To illustrate the propagation of such a plane wave across the simulation domain, the real part of the $E_z$ component is plotted in Figure 4.10. No structure is present in the simulation domain in this case (open space). The wavelength of the incident wave was 1000 nm and the simulation domain was 5 μm by 5 μm (5λ). Straight horizontal wave fronts of the plane wave are clearly recognized.

In the open region configuration we apply a laterally limited source like the linear Gaussian source. With the Gaussian source we can, for instance, represent illumination with a real laser beam, although here it is limited to one lateral direction, as mentioned previously. Similar to the plane wave, the Gaussian source is generated at the top boundary in combination with the absorbing boundary condition. The amplitude of the field is in this case changing according to Gaussian distribution with the intensity peak in the middle of the beam. The mathematical description of the modified boundary condition at the top, according to the applied Gaussian source, is given by Equation (4.6).

$$\frac{\partial E_z}{\partial x} = -j2k_0 E_{inc}(y) + jk_0 E_z - \frac{j}{2k_0}\frac{\partial^2 E_z}{\partial y^2} - \frac{j}{k_m}(2y^2 - 1)E_{inc}(y) \tag{4.6}$$

The symbol $k_m$ corresponds to the wave number of the incident wave and is defined as $k_m = \omega\sqrt{\varepsilon\mu}$, where $\varepsilon$ and $\mu$ correspond to the complex permittivity

---

* Linear because its intensity is changing in only one lateral direction (following the Gaussian function), whereas in another lateral direction there is no change in the 2-D model.

Plane wave source line (top boundary)

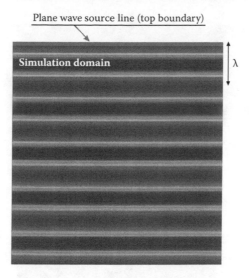

**FIGURE 4.10**    Representation of real part of $E_z$ component of a plane wave applied at the top boundary. At the bottom boundary, absorbing boundary condition was applied. The selection of the boundary condition at the left and right side does not affect the simulation in this case, since the light is propagating in vertical direction only in the entire simulation domain.

and permeability of the incident medium. The $E_{inc}(x)$ is the electric field of the incident wave following the Gaussian function in $x$ direction as given in Equation (4.7). The $\sigma^2$ is the dispersion of the beam in $x$ direction, affecting the width, $w$, of the source.

$$E_{inc}(y) = E_0 e^{-\frac{y^2}{\sigma^2}} \tag{4.7}$$

To keep the propagated light from the source close to the Gaussian form, without excessive dispersion, the width of the initially generated beam should be greater than a few wavelengths. In Figure 4.11 the dispersion level of generated light from the Gaussian source is illustrated by showing the real part of $E_z$ for three different widths of the generated beam. The collimation level improves when the width of the beam is increased. As a rule of thumb, to generate a sufficiently well collimated Gaussian beam, the applied width has to be at least $w = 3\lambda$.

Usually both TE and TM polarizations are present in the applied illumination. Most typically a circular polarization is applied, which consists of 50% of TE and 50% of TM polarization. The optical system is solved for each polarization separately, and then the superposition of the fields is performed to obtain the final solution. In literature, sometimes the simulation results are shown for one polarization only; however, we should be aware that there may be significant

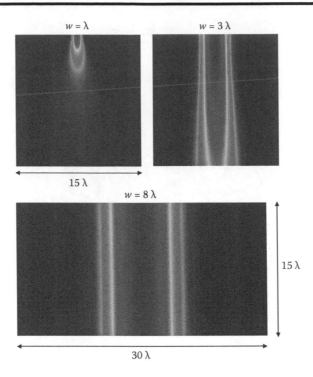

**FIGURE 4.11**   Propagation of Gaussian beam (light intensity) for different ratios of the beam width, defined by $\sigma^2$, and wavelength of light.

differences in results, corresponding to different polarizations. Therefore, both polarizations should be considered despite doubling of the computation time.

### 4.3.5   FORMULATION OF THE SYSTEM OF EQUATIONS

There exist different methods of solving systems of differential equations. In the following we explain an approach based on the Ritz variational method (Ritz 1908) to solve our system of equations, defined by Equation (4.2). In the Ritz variational method the problem is represented by the functional $F$ as given in Equation (4.8)

$$F(\phi) = \frac{1}{2}\langle D\phi, \phi \rangle - \frac{1}{2}\langle \phi, f \rangle - \frac{1}{2}\langle f, \phi \rangle \qquad (4.8)$$

where angular brackets denote the inner product, $\phi$ is the trial function (in our case the electric or magnetic field), $f$ represents the illumination source, and $D$ is the differential operator which in this case is defined by the wave equation

[Equation (4.2)]. The idea behind the variational method is to minimize the functional value with respect to the trial functions. Minima can be found by setting the first derivatives of the functional to zero, as written in Equation (4.9).

$$\frac{\partial F}{\partial \phi_1} = 0, \quad \frac{\partial F}{\partial \phi_2} = 0, \quad \ldots \tag{4.9}$$

Equation (4.9) presents the basis for establishing the set of linear equations describing the system. The system can be formulated for a single element (e) as given in Equation (4.10) and for all elements in the simulation domain as given in Equation (4.11).

$$\left\{ \frac{\partial F^e}{\partial \phi^e} \right\} = K^e \phi - b^e = 0 \tag{4.10}$$

$$\left\{ \frac{\partial F}{\partial \phi} \right\} = \sum_{e=1}^{N} (K^e \phi - b^e) = K\phi - b = 0 \tag{4.11}$$

Matrix $K^e$ corresponds to one element and $K$ corresponds to the whole simulation domain. They consist of coefficients defining relations between the field in different nodes. This relation is basically defined by wave equations and by the node position and the surrounding material. Using linear approximation [Equation (4.3)] for the field, the size of the matrix $K^e$ is $3 \times 3$ (triangular element has three nodes), whereas the size of the matrix $K$ is $N \times N$ where $N$ is the number of nodes in the simulation domain. In our structures $K$ is not a positive definite sparse symmetric matrix; the elements are complex numbers. Vector $\phi$ in Equations (4.10) and (4.11) represents the field values at each node, whereas vector $b$ includes information about the field of illumination at the nodes located at the top boundary.

## 4.3.6   SOLVING THE LINEAR SYSTEM OF EQUATIONS

The system is generally defined by Equation (4.11). We have to be aware that the system can be relatively large if dense discretization of the simulation domain is applied (millions of unknowns; one node is one unknown). Thus, we have to employ special methods for solving large sparse systems. Direct methods are not favorable in this case because most of the elements in the large matrix $K$ are equal to zero, while direct methods are performing fill-ins into the matrix using memory locations of the computer. The faster methods that only need to allocate computer memory for non-zero elements, a few memory locations to indicate the elements position in the matrix, and a few additional vectors of size $N$ to memorize the values of elements are iterative methods (Barrett et al. 1987; Saad 2003). A compact memory format to efficiently store the elements of sparse matrix is preferable for our system (Press et al. 2002).

Among the iterative methods the stationary methods are not very suitable since the matrix $K$ is not positive definite and thus the convergence with the stationary iterative methods, such as the Jacobian, Gaus-Seidl, successive overrelaxation (SOR), and symmetric successive overrelaxation (SSOR) methods, cannot be obtained.

The methods that are more applicable to our system are gradient methods (Barrett et al. 1987), such as a special form of conjugate gradient method (CG) (Jin 2002), biconjugate gradient method (BiCG), conjugate gradient squared (CGS) (Barrett et al. 1987; Saad 2003), and special forms of BiCG methods such as biconjugate gradient stabilized (BiCGSTAB) (Barrett et al. 1987) or BiCGSTAB(l) (Sleijpen and Fokkema 1993).

According to our tests the best results for thin-film solar cell structures can be obtained with the BiCG method, which can be simplified significantly, as the matrix $K$ is symmetrical. To additionally improve convergence and also to stabilize the method, an upgrade with Jacobian and SSOR preconditioner is preferable (Barrett et al. 1987; Saad 2003).

### 4.3.7  DETERMINATION OF THE OUTPUT PARAMETERS OF THE SIMULATION

After the system of equation is solved, the complex values of the fields $E_z$, in case of TE, and $H_z$, in case of TM polarization, at each node in the simulation domain are obtained. The raw optical data have to be transformed further into standard solar cell output parameters. As the main output parameters related to optical analysis we consider here:

- total reflectance from the structure,
- absorptance in the individual layers,
- quantum efficiency and short-circuit current calculated from the absorptance in the absorber layer, and
- the 2-D generation rate across the structure of the photo-generated electrons and holes in active layers.

These parameters are obtained by post-processing of electric and magnetic field raw data.

The total reflectance $R_{tot}$ is calculated at the top boundary of the simulation domain, where the incidence wave is generated and defined across the entire boundary. We calculate $R_{tot}$ based on the Poynting vector $S = ½ (E \times H^*)$, representing the power density (flux) of electromagnetic waves. Here, the electric and magnetic field correspond to the same polarization. As explained in previous chapters, $H$ in the case of TE polarization and $E$ in the case of TM polarization can always be represented by $E$ from TE and $H$ from TM polarization, respectively.

$R_{tot}$ is defined and calculated as the ratio between the reflected light intensity and the incident intensity. For both TE and TM polarization, final expressions for reflectance are given by Equation (4.12) (Čampa 2009):

$$R_{TE} = \left| \frac{Re(S_{y\_ref})}{Re(S_{y\_inc})} \right| = \left| \frac{Im(\frac{E_{ref}}{\mu^*} \frac{\partial E_{ref}^*}{\partial x})}{Re(\frac{k_m}{\mu^*} E_{inc} E_{inc}^*)} \right|$$

$$R_{TM} = \left| \frac{Im(\frac{H_{ref}^*}{\varepsilon} \frac{\partial H_{ref}}{\partial x})}{Re(\frac{k_m}{\varepsilon} H_{inc} H_{inc}^*)} \right| \qquad (4.12)$$

where the subscript "inc" stands for the incident and "ref" for reflected field.

Absorptance in an individual layer, $A_{layer}$, which is defined as the sum of absorptances in all discrete elements forming the layer, can be calculated using the final Equation (4.13) (Čampa 2009). Again, the Poynting vector was used to derive the final equation:

$$A_{layer} = \sum_{\substack{layer \\ volume}} A_{layer}^e; \quad A_{layer}^e = \frac{\omega \varepsilon_0 \, \varepsilon_{layer\_Im}^e}{2} EE^* + \frac{\omega \mu_0 \mu_{layer\_Im}^e}{2} HH^* \qquad (4.13)$$

where $\varepsilon_{layer\_Im}^e$ and $\mu_{layerl\_Im}^e$ are imaginary parts of the permeability and permittivity corresponding to the layer. Wavelength dependencies of quantities are not denoted in these equations.

Another important optical parameter is the 2-D generation rate, $G_L$, of the photo-generated electron and hole pairs in the active layers. The $G_L$ represents the main input parameter for further electrical simulations of photovoltaic devices. At a specific point in the active layer, the $G_L^e$ is calculated from the absorptance $A_{layer}^e$ as given in Equation (4.14):

$$G_L^e(\lambda_i) = \frac{\lambda_i}{hc} \frac{2}{l_y} A_{layer}^e(\lambda_i) \cdot I_{inc}(\lambda_i) \qquad (4.14)$$

where $h$ is Planck's constant ($h = 6.626 \; 10^{-34}$ Js), $c$ is the velocity of light in free space ($c = 299792458$ m/s), $I_{inc}$ is illumination power density (W/m$^2$) at a specific discrete wavelength $\lambda_i$, and $l_y$ is the discrete element length in the $y$-direction.

As we know from 1-D modeling, short-circuit current density, $J_{SC}$, and external quantum efficiency, $QE$, correspond to opto-electronic and not only optical properties of photovoltaic devices. However, considering the same assumptions as in the 1-D optical analysis (ideal extraction of all optically generated electrons and holes in active layers, neglecting contributions from supporting layers such as $p$- and $n$-doped in the case of thin-film silicon solar cells), $J_{SC}$ and $QE$ can be determined directly from the results of optical simulation: $QE = A_{\text{active\_layer}}$ and $J_{\text{SC}}$ as given in Equation (4.15)

$$J_{\text{SC}} = \sum_{i=1}^{M} \frac{q\,\lambda_i}{h\,c} A_{\text{abs}}(\lambda_i) \cdot I_{\text{inc}}(\lambda_i) \qquad (4.15)$$

where $q$ is the elementary charge ($q = 1.602\ 10^{-19}$ As) and $M$ is the number of discrete wavelength intervals of the applied illumination spectrum.

As we explained, usually both TE and TM polarizations are present in the applied illumination. Since the fields, either $E$ or $H$, corresponding to TE and TM are perpendicular to each other, the superposition of these two types of illuminations can be considered also on the light intensity levels. This means also that all the mentioned output parameters ($R_{\text{tot}}$, $A_{\text{layer}}$, $G_L$) can be calculated separately for TE and TM polarization and then summed up.

### 4.3.8   OPTICAL SIMULATOR FEMOS 2-D

Based on the presented optical model, a computer simulator FEMOS 2-D (two-dimensional finite element method optical simulator) was developed (Čampa, Krč, and Topič 2008). Here we present it as an example of 2-D optical simulators based on FEM that was specifically developed for analysis of thin-film photovoltaic devices. Different simulators based on the FEM method (e.g., Ansys HFSS, COMSOL, and JCMWave) are available for general use to solve electromagnetic, mechanical, and other types of problems.

FEMOS 2-D comprises a mathematical kernel, user interface, and module for graphical representation of output results. The computer code for all three modules was written in C++ language.

The user-friendly interface (Figure 4.12) facilitates

- the definition of structure and optical properties of the analyzed device (complex refractive indexes and thicknesses of individual layers, definition of interface textures),
- the configuration of the simulation domain and its parameters (illumination, boundary conditions), and
- the definition of numerical settings (selection of different numerical methods) and loading and saving complete projects.

In the mathematical kernel of the FEMOS 2-D simulator, we implemented the model described in the previous sections. We paid special attention to the optimization of the C++ code, since the calculation speed is extremely crucial and represents the bottleneck of such simulators. Additionally, the code was compiled with the GCC GNU compiler (GCC GNU), which can take the full advantage of the latest instruction sets in the modern x86- or x86-x64-based processors.

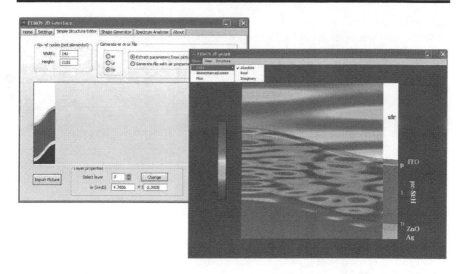

**FIGURE 4.12**   The picture of FEMOS 2-D user interface (top left) and one of the windows for presentation of the calculated results (bottom right).

The graphical presentation module of the simulator transforms the calculated data to a user-friendly presentation of results. In Figure 4.12 an example of a picture from the graphical presentation module is shown. The user can present the values of the electric or magnetic field (absolute value, real or imaginary part), light intensity distribution, and absorptance in a specific layer in the structure.

### 4.3.9   VERIFICATION OF THE SIMULATOR ON REALISTIC STRUCTURES

The simulator was verified on different realistic thin-film structures. Here, we present results of verification on two simple structures. The first structure is a 1-D grating reflector with rounded texture close to a sinusoidal shape (Figure 4.13). The texture was made by embossing the polycarbonate substrate

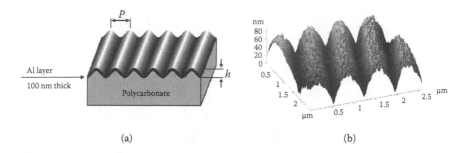

(a)                                           (b)

**FIGURE 4.13**   Analyzed structure with 1-D sinusoidal surface texture (a) and AFM scan on the surface of the Al layer (b).

(methods used in CD, DVD, or blue-ray disc production) by the company OM&T from the Netherlands. On top of the grating, a layer of 100-nm-thick aluminum alloy was deposited.

To determine the surface texture of the sample, we performed atomic force microscopy measurements (Figure 4.13b). A rounded shape, close to sinus, was detected for the tops of the texture features. Sharper shape of the valleys can be observed in the AFM picture. Partially this is assigned to an artifact of AFM measurements and partially the not ideally sinusoidal shape of the actual surface texture of the Al covering layer. Regarding the AFM measurements it turned out that the type of the AFM tip affects the detected shape of the valleys in AFM scans. Detection of narrow valleys becomes even more problematic in rectangular textures with small periods and large heights. Special sharp AFM tips need to be used in such cases. Despite these deviations between the detected shape and ideal sine shape, we consider sinusoidal approximation in our simulations. The period and the height of the applied texture were $P \approx 700$ nm and $h \approx 40$ nm as obtained from the measurements.

In Figure 4.14 the simulation domain and the discretization (inset) of the structure are shown. The polycarbonate substrate was not included in simulation, since the Al layer with the thickness of 100 nm does not transmit any light to the substrate.

The simulation domain was built up from 140,000 triangular elements, resulting in an excellent approximation of the sinusoidal shape and ensuring enough points inside the layer following the rule to satisfy the conditions regarding the effective wavelength. The sinusoidal shape agrees quite well with

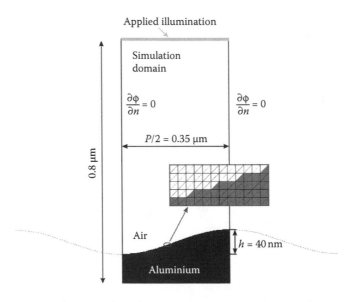

**FIGURE 4.14**    Definition of the simulation domain and an inside view into discretization of the analyzed structure.

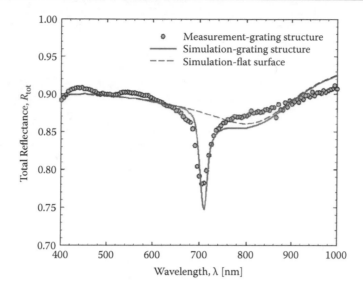

**FIGURE 4.15** Measured and simulated total reflectance from the Al surface of the analyzed structure (sinusoidal texture with *P* = 700 nm and *h* = 40 nm).

the measured shape; however, we could also import a realistic shape of the profile into a simulator. The other input parameters of simulation were measured wavelength-dependent refractive indices of the Al alloy and the polarization of light. In simulations we used 50% of TE and 50% of TM polarization, approaching the realistic illumination in our measurements.

The measurements of total reflectance were performed by Lambda 950 spectrometer with an integrating sphere. In Figure 4.15 the results of measurement and simulations are compared. Good agreement is observed for the simulated curve considering the surface texturization. As a reference also a simulation of the structure with a flat surface is shown. It is evident that the dip in the $R_{tot}$ characteristics at $\lambda \approx 700$ nm corresponds to surface texturization. At this wavelength the first diffraction order takes place, and a part of the light is scattered into a large angle. The simulations showed that the dip in $R_{tot}$ is related to additional absorption inside the Al alloy, since this wavelength of light is intensively scattered in large angles.

As the second verification example we present simulations of a structure with a thin i-a-Si:H layer (200 nm) on top of an Al alloy layer (100 nm), both deposited on a periodically textured polycarbonate substrate. The a-Si:H layer was deposited by PECVD at Technical University Delft (TU Delft), the Netherlands. The substrate had a somewhat different texture than in the previous case: the shape of the texturization features was rectangular-like (trapezoidal, inclination angles of the vertical walls 80 degrees) and the period and height were $P = 300$ nm and $h = 40$ nm, respectively. The cross-section of the

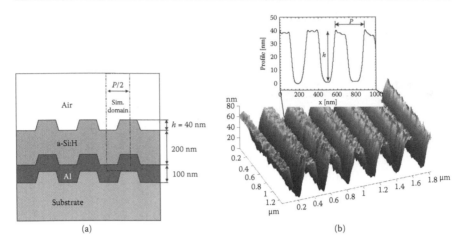

(a)                                                            (b)

**FIGURE 4.16**   Schematic representation of the analyzed structure with a-Si:H layer (a) and AFM scan on the top a-Si:H surface (b).

simulated structure and the AFM measurements on the i-a-Si:H top surface on a real device are shown in Figure 4.16.

All of these parameters were obtained from AFM and were considered for the structure definition in our simulations. Considering the periodic nature of gratings and symmetry of rising and falling slope, the periodic-symmetric boundary condition was used in simulations, and only half of the period is taken into simulation domain. The reader can guess which boundary conditions have to be used. The entire simulation domain consisted of approximately 70,000 triangular elements, which is half of the elements that were needed in the previous example. Even though the structure is thicker, the number of elements also depends on the period, which is now lower (300 nm instead of 700 nm in the previous case). The number of elements is sufficient for a good stepless approximation of all grating details.

In Figure 4.17 the simulated and measured $R_{tot}$ are shown, and good agreement is observed in the position of all of the interference fringes that occur due to grating and due to the thickness effect in this case. Besides the agreement in the position of the interference fringes, the amplitudes of fringes are sufficiently reproduced in simulation. To distinguish between the thickness interference fringes and grating interference fringes, we performed additional simulation on the same structure but considering flat interfaces. The additional interferences that originate from the grating effect are the consequence of light scattering at the first (air/i-a-Si:H) or last (i-a-Si:H/Al) interface. The efficient scattering of light, especially to large scattering angles, results in increased absorptance in the layers, leading to decreased $R_{tot}$ at certain wavelengths. In the case of gratings the scattering angles are discrete and they change with the wavelength of light, according to the grating equation (Heine and Morf 1995). Simulations show the interference

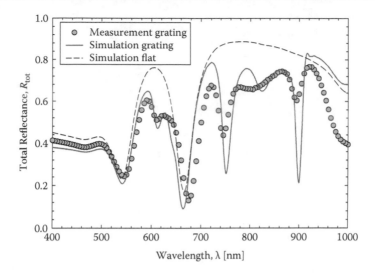

**FIGURE 4.17**   Measured and simulated total reflectance of the structure shown in Figure 4.16.

(decrease in $R_{tot}$) at the wavelength of 600 nm is the consequence of air/i-a-Si:H interface. The interference corresponds to the efficient light scattering of the second scattering order to a large discrete angle (close to 90 degrees) inside the i-a-Si:H layer. Other interference decreases corresponding to the grating effect arise mostly because of scattering at back (i-a-Si:H/Al) interface, as indicated by simulations.

The deviations that can be observed generally between the measured and simulated results on presented structures, as well as the structures of complete solar cells we refer to, include the following:

- less than perfect conformal growth of all thin layers on textured substrates (not ideal transfer of the texture from bottom to top interfaces),
- some uncertainties in refractive indices of layers,
- non-homogeneous properties of layers,
- imperfectly reproduced shapes of grating interfaces in simulations, and
- a possible difference between the properties of light used in measurements and simulated monochromatic light of a specific wavelength.

Taking these and other verification results into account, we consider the FEMOS 2-D simulator to be a credible tool for further investigation of thin-film solar cell structures.

The advantage of your own developed simulator is that you have full access to everything in the code and you can play and try to optimize your simulator in every detail. You can display each quantity you wish in a desired format. However, developing simulators takes time (especially debugging); therefore, it is always recommended to check what simulators are available on the market, what their functionality is, and how much they cost.

# REFERENCES

Ansys HFSS. http://www.ansys.com/Products/Simulation+Technology/Electromagnetics/ High-Performance+Electronic+Design/ANSYS+HFSS.

Atwater, H. A., and A. Polman. 2010. "Plasmonics for Improved Photovoltaic Devices." *Nature Materials* 9 (3): 205–213. doi:10.1038/nmat2629.

Barrett, R., M. Berry, T. F. Chan, J. Demmel, J. Donato, J. Dongarra, V. Eijkhout, R. Pozo, C. Romine, and H. van der Vorst. 1987. *Templates for the Solution of Linear Systems: Building Blocks for Iterative Methods.* Society for Industrial and Applied Mathematics.

Bittkau, K., and T. Beckers. 2010. "Near-Field Study of Light Scattering at Rough Interfaces of a-Si:H/mu c-Si:H Tandem Solar Cells." *Physica Status Solidi a–Applications and Materials Science* 207 (3) (March): 661–666. doi:10.1002/pssa.200982671.

Bittkau, K., M. Schulte, M. Klein, T. Beckers, and R. Carius. 2011. "Modeling of Light Scattering Properties from Surface Profile in Thin-Film Solar Cells by Fourier Transform Techniques." *Thin Solid Films* 519 (19) (July 29): 6538–6543. doi:10.1016/j.tsf.2011.04.122.

Born, M., and E. Wolf. 1999. *Principles of Optics.* 7th ed. Cambridge University Press.

Čampa, A. 2009. *Modelling and Optimisation of Advanced Optical Concepts in Thin-Film Solar Cells.* PhD thesis, University of Ljubljana.

Čampa, A., J. Krč, F. Smole, and M. Topič. 2008. "Potential of Diffraction Gratings for Implementation as a Metal Back Reflector in Thin-Film Silicon Solar Cells." *Thin Solid Films* 516 (20): 6963–6967. doi:10.1016/j.tsf.2007.12.051.

Čampa, A., J. Krč, and M. Topič. 2008. "Two-Dimensional Optical Model for Simulating Periodic Optical Structures in Thin-Film Solar Cells." *Informacije MIDEM* 38 (1): 5–10.

Chang, S. T., B.-F. Hsieh, and Y. C. Liu. 2012. "A Simulation Study of Thin Film Tandem Solar Cells with a Nanoplate Absorber Bottom Cell." *Thin Solid Films* 520 (8) (February 1): 3369–3373. doi:10.1016/j.tsf.2011.10.094.

COMSOL. http://www.comsol.com/.

Dewan, R., I. Vasilev, V. Jovanov, and D. Knipp. 2011. "Optical Enhancement and Losses of Pyramid Textured Thin-Film Silicon Solar Cells." *Journal of Applied Physics* 110 (1) (July 1). doi:10.1063/1.3602092.

Ferry, V. E., A. Polman, and H. A. Atwater. 2011. "Modeling Light Trapping in Nanostructured Solar Cells." *Acs Nano* 5 (12) (December): 10055–10064. doi:10.1021/nn203906t.

GCC GNU. http://gss.gn.org.

Haase, C., and H. Stiebig. 2007. "Thin-Film Silicon Solar Cells with Efficient Periodic Light Trapping Texture." *Applied Physics Letters* 91: 061116. doi:10.1063/1.2768882.

Heine, C., and R. Morf. 1995. "Submicrometer Gratings for Solar-Energy Applications." *Applied Optics* 34 (14) (May 10): 2476–2482.

Isabella, O., H. Sai, M. Kondo, M. Zeman. 2012a. "Full-wave optoelectrical modeling of optimized flattened light-scattering substrate for high efficiency thin-film silicon solar cells." *Prog. Photovolt: Res. Appl.* Wiley. doi: 10.1002/pip.2314.

Isabella, O., S. Solntsev, D. Caratelli, and M. Zeman. 2012b. "3-D Optical Modeling of Thin-Film Silicon Solar Cells on Diffraction Gratings." *Progress in Photovoltaics: Research and Applications.* doi:10.1002/pip.1257. http://onlinelibrary.wiley.com/doi/10.1002/pip.1257/abstract.

Jandl, C., W. Dewald, C. Pflaum, and H. Stiebig. 2010. "Simulation of Microcrystalline Thin-Film Silicon Solar Cells with Integrated AFM Scans." In *Proceedings of the Twenty-Fifth European Photovoltaic Solar Energy Conference and Exhibition,* 3154–3157. doi:10.4229/25thEUPVSEC2010-3AV.2.3

Jandl, C., K. Hertel, and C. Pflaum. 2011. "Simulation of Thin-Film Silicon Solar Cells with Integrated AFM Scans for Oblique Incident Waves." In *Proceedings of the Twenty-Sixth European Photovoltaic Solar Energy Conference and Exhibition,* 2663–2666. doi:10.4229/26thEUPVSEC2011-3AV.2.23

JCMWave. http://www.jcmwave.com/.

Jin, J.-M. 2002. *The Finite Element Method in Electromagnetics.* 2nd ed. Wiley-IEEE Press.

Kapadia, R., Z. Fan, K. Takei, and A. Javey. 2012. "Nanopillar Photovoltaics: Materials, Processes, and Devices." *Nano Energy* 1 (1) (January): 132–144. doi:10.1016/j.nanoen.2011.11.002.

Kong, J. A. 1990. *Electromagnetic Wave Theory.* 2nd ed. Wiley-Interscience.

Lipovšek, B., J. Krč, and M. Topič. 2011. "Optical Model for Thin-Film Photovoltaic Devices with Large Surface Textures at the Front Side." *Informacije MIDEM* 41 (4).

Lockau, D. 2012. *Optical Modeling of Thin-Film Silicon Solar Cells with Random and Periodic Light Management Textures.* PhD thesis, Technical University, Berlin.

MEEP simulator. http://ab-initio.mit.edu/wiki/index.php/Meep.

Press, W. H., S. A. Teukolsky, W. T. Vetterling, and B. P. Flannery. 2002. *Numerical Recipes in C++: The Art of Scientific Computing.* 2nd ed. Cambridge University Press.

Ritz, W. 1908. "Ueber eine neue Methode zur Lösung gewisser Variationsprobleme der mathematischen Physik." *Journal für die reine angewandte Mathematik* 1908 (135): 1–61.

Rockstuhl, C., S. Fahr, K. Bittkau, T. Beckers, R. Carius, F.-J. Haug, T. Söderström, C. Ballif, and F. Lederer. 2010. "Comparison and Optimization of Randomly Textured Surfaces in Thin-Film Solar Cells." *Optics Express* 18 (S3): A335–A341. doi:10.1364/OE.18.00A335.

Saad, Y. 2003. *Iterative Methods for Sparse Linear Systems.* 2nd ed. Society for Industrial and Applied Mathematics.

Sleijpen, G. L. G., and D. R. Fokkema. 1993. *Electronic Transactions on Numerical Analysis.* 1, (September): 11–32.

Springer, J., A. Poruba, and M. Vaneček. 2004. "Improved Three-Dimensional Optical Model for Thin-Film Silicon Solar Cells." *Journal of Applied Physics* 96 (9) (November 1): 5329–5337. doi:10.1063/1.1784555.

# Applications

# One-Dimensional Optical Simulations

5

## 5.1   INTRODUCTION

In previous chapters we dealt with the development of 1-D and 2-D optical models for simulations of thin-film photovoltaic devices. In this chapter we present six case studies of application of the 1-D optical modeling and simulations, investigating different optical concepts and related effects in thin-film solar cells. The presented concepts aim to enhance light absorption in absorber layers and minimize optical losses in the supporting layers, leading to improved conversion efficiencies of solar cells.

In the first case we present the study of enhanced light scattering that can be introduced in thin-film silicon solar cells by highly textured transparent conductive oxide (TCO) superstrates. In the second case we demonstrate the use of modeling and simulations in the absorber thickness determination in tandem and triple-junction solar cells. In the third case we focus on investigation and optimization of intermediate reflector in tandem micromorph silicon solar cells. In the fourth case study we employ modeling and simulations to design antireflective coatings (ARCs). The fifth one deals with 1-D photonic crystals as structures for light manipulation in solar cells. With the sixth case study we show an approach to 1-D optical modeling of white dielectric back reflectors, in particular white paint reflectors in thin-film solar cells.

As pointed out in previous chapters, besides having reliable optical models we have to determine and use realistic input parameters to perform reliable simulations. In the 1-D simulations presented here, we employ the Sun*Shine* optical simulator (see its description in Chapter 2); however, other simulators with similar features could be used. Characterization techniques for determination of input parameters presented in Chapter 3 can be used to determine the input parameters of simulations.

## 5.2 POTENTIAL OF ENHANCED LIGHT SCATTERING IN THIN-FILM SOLAR CELLS

Light scattering at textured interfaces is an important technique of light management in thin-film silicon solar cells. Scattered light rays experience prolonged optical paths in thin semiconductor (absorber) layers, due to non-perpendicular propagation (effect 1 in Figure 5.1).

At the back side of the solar cell, the light that was not fully absorbed within its first pass throughout the structure (long-wavelength light) has to be efficiently reflected back (effect 2 in Figure 5.1). Additional scattering at the back reflector is usually present due to surface texture of the reflector (the texture of the TCO superstrate is more or less transferred to all interfaces in thin-film solar cell structures). Reflected rays that are not fully absorbed on their way back throughout the absorber can experience another reflection at the front interfaces, which appears to be crucial for light trapping in solar cell structures (effect 3 in Figure 5.1). The rays that propagate in angles equal to or larger than the angle of total reflection of the p/TCO interface are totally reflected back and thus trapped in semiconductor layers. (Reflectance close to 100% can be achieved if the layers are not highly absorbing in the wavelength region of interest.) The angle of total reflection is defined according to Snell's law as

$$\varphi_{tot} = \text{Re}\left[ \arcsin\left( \frac{N_2}{N_1} \right) \right] \tag{5.1}$$

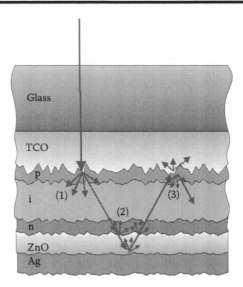

**FIGURE 5.1**   A schematic of light scattering and trapping in a thin-film silicon solar cell (deposited on a textured TCO superstrate); three distinctive effects (1)-(3) are denoted.

where $N_1$ and $N_2$ are complex refractive indexes of the $p$-doped Si layer and front TCO contact, respectively. Because the refractive indexes of Si layers are larger ($n = 3-4$) than those of TCOs ($n \approx 2$), the wave-guiding effect inside semiconductor layers (absorbers) is possible.

In thin-film silicon or other types of solar cells with indirect semiconductor absorbers, light trapping techniques are very important because of the low absorption coefficient of the material around the optical bandgap. In the superstrate type of solar cells, important characteristics of front TCO superstrates present their scattering characteristics, since they introduce light scattering into the cell structure. Many groups have been developing and optimizing the texture of the TCO surface to achieve efficient light scattering, especially in the long-wavelength part of the spectrum where the absorption is low (Springer, Poruba, and Vaneček 2004). If a microcrystalline silicon absorber (i-μc-Si:H) is included in the structure instead of or besides the amorphous absorber (i-a-Si:H), light trapping becomes even more important. In this case the light scattering abilities of textured surfaces have to be extended to the wavelength of 1200 nm. Therefore, besides the TCOs that were developed and optimized for cells with amorphous absorbers, such as the Asahi U-type TCO, new solutions have been investigated. Different types of $SnO_2$:F-, ZnO:B-, and ZnO:Al-based TCO with random nano-textures with pyramidal and crater-like features have been developed for μc-Si:H and a-Si:H solar cells (Kluth et al. 1999;

Fay et al. 2007). In addition to these solutions, realizations of more advanced surface textures, such as double-textured TCOs, have been developed (Kambe et al. 2003; Oyama et al. 2008). These TCOs facilitate a high scattering level (haze parameter) also in the long-wavelength part of the spectrum. However, further improvements in scattering properties of TCO superstrates are of interest. Modeling and simulations can be used to show how much we can gain with the superstrates with a high haze parameter and justify and give directions for further improvements. Besides high haze, we demonstrate how important the angular distribution function ($ADF$) of scattered light is.

One-dimensional optical modeling and simulations have been proven to be effective tools for investigation of optical situation in thin-film solar cells with nano-textured interfaces. In the next section we use them to investigate the effects of enhanced light scattering on the performance of thin-film silicon solar cells. Simulations enable us to investigate the role of both scattering parameters—the haze parameter and the angular distribution function—together or decoupled. As mentioned, both are linked to the morphology of textured interfaces. In this analysis we treat them as independent input parameters in order to get a clear picture of the impact of each.

## 5.2.1    THE EFFECT OF HIGH HAZE AND BROAD ANGULAR DISTRIBUTION FUNCTION IN SINGLE-JUNCTION Si SOLAR CELLS

We investigate the effects of enhanced scattering properties of textured TCOs on a μc-Si:H solar cell in a superstrate configuration, with the basic structure presented in Figure 5.1 (details are given later).

First, we apply different wavelength-dependent haze parameters, as presented in Figure 5.2. What impact will this have on the optical performance of the cell?

The curves in the figure correspond to the haze parameter in transmission for the TCO/air interface. For the TCO layer we considered $SnO_2$:F material in all cases of this analysis. As a reference haze the measured haze of the Asahi U-type $SnO_2$:F TCO superstrate was considered (Krč et al. 2002) (see the morphology of the surface in Chapter 3, Figure 3.10b). To study the effects of enhanced haze parameters on solar cell performance, hypothetical high haze TCOs (HHTs) were introduced in simulations. Additional value in the name of HHTxx represents the haze value (in %) at the wavelength of 550 nm. Practically, such high haze levels can be achieved by TCO superstrates with advanced surface textures such as double textured $SnO_2$:F or ZnO:Al TCO (Kambe et al. 2003; Oyama et al. 2008; Hongsingthong et al. 2010).

In addition to varying the haze parameter, we vary and investigate also the role of angular distribution of scattered light. The $ADF$s that are included in the analysis are presented in Figure 5.3. They correspond to $ADF$s of transmitted

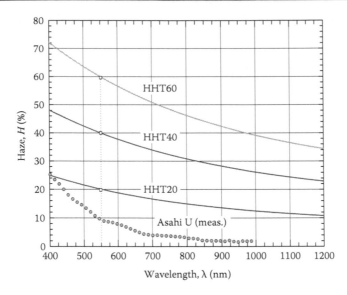

**FIGURE 5.2**    Haze parameters of transmitted light at the TCO/air interface used in our simulations.

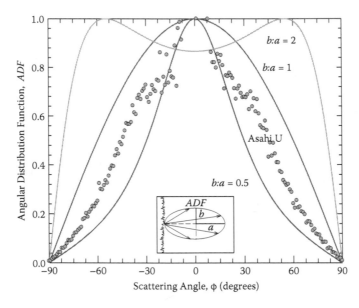

**FIGURE 5.3**    Angular distribution function (*ADF*) of scattered light used in simulations. *ADF*s with the shape of geometrical ellipsis were chosen for the analysis. The radius ratio b:a defines the broadness of the ellipsis and thus the broadness of the *ADF*. If b:a is greater than 1, the peak at 0 degrees starts to split into two peaks, shifting toward larger angles (see example with b:a = 2). Note that the curves are plotted in a Cartesian plot; in a polar plot they would have the shape of the ellipsis, as shown in the inset.

scattered light at the $SnO_2$:F/air interface. One-dimensional representation of the *ADF*s is used (as measured in the plane perpendicular to the sample; no 3-D transformation is included in the figure. For the transformation see Chapter 3, Section 3.2.6.3).

As a reference we consider the Asahi U-type superstrate. Its measured *ADF* characteristic at $\lambda = 633$ nm is plotted in Figure 5.3 by circles. To study the role of the *ADF* on the performance of µc-Si:H solar cells, we introduced different hypothetical *ADF*s in our simulations. We chose the shape of geometrical ellipsis (see inset in Figure 5.3) for the hypothetical *ADF*s. Using ellipses the *ADF* can be modified in a simple way, just by changing the ratio of radius *b:a* of the ellipsis. On the other hand, ellipses were found as a good approximation of *ADF*s of some realistic TCO superstrates (e.g., etched ZnO:Al TCO). The larger the radius ratio *b:a* is, the broader is the *ADF* of scattered light (more light is scattered into large scattering angles). Ellipsis with *b:a* = 1 represent the Lambertian (cosine function resulting in a circular shape) distribution function.

We investigate the effects of enhanced scattering on quantum efficiency and short-circuit current density of the µc-Si:H solar cell in the following configuration: glass (1 mm)/$SnO_2$:F (800 nm)/p-µc-Si:H (20 nm)/i-µc-Si:H (500 nm)/n-µc-Si:H (20 nm)/ZnO (100 nm)/Ag. The cell has a relatively thin i-µc-Si:H absorber layer, so it is expected that the light scattering effects will have higher influence than in thick absorbers. It is the general trend to reduce absorber thickness.

Based on the haze parameter for transmitted light at the TCO/air interface presented in Figure 5.2, we calculated the corresponding haze parameters for both transmitted and reflected light for each internal textured interface. Scalar scattering theory was used in these calculations (see Chapter 3), considering that the morphology of the textured TCO surface is transferred to all succeeding interfaces with the loss in $\sigma_{rms}$ of 3 nm per 100 nm of the layer thickness (TCO/p, p/i, i/n, n/Ag). In these simulations we apply the same *ADF* to all textured internal interfaces for reflected and transmitted light (as presented in Figure 5.3, applying 3-D transformation). Wavelength independency of the *ADF*s was also assumed.

In Figure 5.4 the effects of haze variation on *QE* of the µc-Si:H solar cell are shown. The *ADF* of the Asahi U-type TCO was applied in all cases. Fully coherent incident light was assumed. The corresponding $J_{SC}$ values of the cells are given in the figure. Simulation of a cell with ideally flat interfaces is shown in addition, to indicate the improvements related to light scattering with respect to the cell with no scattering. According to the presented results, significant improvement in *QE* and $J_{SC}$ is assigned to the scattering, comparing the *QE* curve and $J_{SC}$ value of the flat cell with others. However, the flat cell does not fully correspond to the µc-Si:H cell deposited on flat superstrate, since some roughness is introduced to the interfaces at the back side as a result of crystalline growth of the µc-Si:H layers.

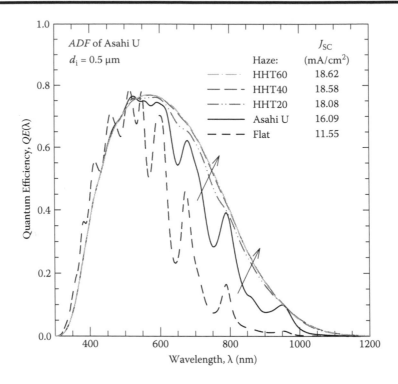

**FIGURE 5.4**    Simulated *QE* curves of the μc-Si:H solar cell on the superstrate with a different haze parameter and the same *ADF* of the Asahi U superstrate.

Next, we compare the cells with HHT superstrates with the Asahi U-type TCO superstrate. Applying HHT20 or HHT40, noticeable improvement in the long-wavelength part ($\lambda > 550$ nm) of the *QE* can be observed (Figure 5.4). This leads to improvement of $J_{SC}$ for more than 15% for HHT40 compared to the cell on Asahi U-type TCO. Also, the interference fringes are smoothed in the *QE* curves, indicating that besides short-wavelength light most of the long-wavelength light is scattered. Another indication about the high level of scattered light is the saturation observed in *QE* and $J_{SC}$ values, comparing the results for HHT40 and HHT60. Increasing the haze above that of HHT40, no important further gain is expected, according to simulations.

Let us proceed to the investigation of the effects of *ADF* variations now. Simulation results are shown in Figure 5.5. The HHT40 haze parameter was applied in all simulations in this case. *ADF*s of elliptical shape with different *b:a* ratios, as presented in Figure 5.3, were used. We can observe significant improvements in the long-wavelength *QE* related to broad *ADF*s ($b:a \geq 1$), compared to the *ADF* of Asahi U-type TCO. However, a small decrease of *QE* is indicated in the short-wavelength part ($\lambda < 550$ nm), which can be explained by

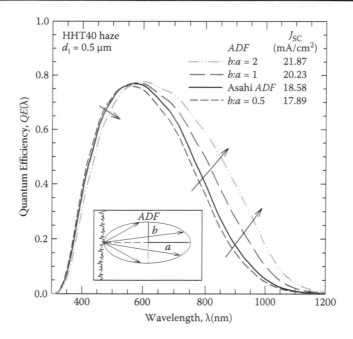

**FIGURE 5.5**  Simulated *QE* curves of a μc-Si:H solar cell on the superstrate with a different *ADF* parameter and the same haze HHT40.

increased optical absorption in p-μc-Si:H layer, since light is scattered already at the TCO/p interface. Therefore, optical paths are prolonged already in the p-μc-Si:H layer and less light enters the i-μc-Si:H absorber layer. Although this effect is present also for long-wavelength light, it is not so evident due to lower absorption coefficient at these wavelengths. The improvements in $J_{SC}$ of the cells are also summarized in the figure. Comparing the $J_{SC}$ of the cell with the *ADF* of Asahi U, 8.8% and 17.7% higher $J_{SC}$ is achieved if broader elliptical *ADF* with $b{:}a = 1$ and 2 is applied, respectively. We should note that these improvements are added to the ones obtained with enhanced haze parameter.

In Figure 5.6 a graphical representation of the effects of *ADF* and haze parameter on $J_{SC}$ is given. The levels of $J_{SC}$ corresponding to the cell with the *ADF* of Asahi U TCO and two different haze levels (Asahi U and HHT40) are plotted as a reference by dashed horizontal lines. A noticeable increase in $J_{SC}$ is observed when the *ADF* is broadened. Additional simulations revealed that saturation in $J_{SC}$ with respect to *ADF* occurs at extremely broad *ADF* with $b{:}a \approx 4$ (not shown here), which would be difficult to realize.

Presented simulations in this section show that besides improving the haze parameter, it is very important to ensure a broad *ADF* of the textured TCO superstrates. Finally, a combination of both high haze and broad *ADF* leads to enhanced performance of thin-film solar cells. By using optical modeling

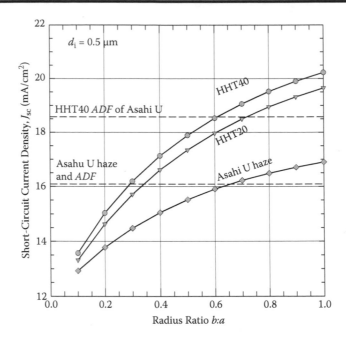

**FIGURE 5.6**   Simulated $J_{SC}$ as a function of *ADF* radius ratio, *b:a*, for different haze parameters.

and simulations calibrated to our cells, we can predict these improvements quantitatively.

## 5.2.2   THICKNESS REDUCTIONS

Enhanced scattering properties can be used to reduce the thickness of the absorber layer by keeping the $J_{SC}$ at a sufficiently high level. We employed optical simulations to study the potential thickness reductions in the µc-Si:H solar cell. We did a set of simulations of a solar cell with various thicknesses of the absorber layer. In Figure 5.7 the summary of the results is presented. The $J_{SC}$ as a function of the i-µc-Si:H absorber layer thickness, by applying different scattering parameters of textured interfaces, is plotted. Almost logarithmic dependence (linear in semi-logarithmic plot) is obtained. From the graph we can observe that, for example, a 0.6-µm thick µc-Si:H solar cell with a HHT40 haze parameter and elliptical *ADF* with $b:a = 1$ can render almost the same $J_{SC}$ as a 2-µm thick solar cell with Asahi U haze parameter and Asahi U *ADF*, or a solar cell with perfectly flat interfaces and $d_i$ of 5 µm. The presented simulation results indicate a relatively high potential for thickness reductions in thin-film Si solar cells if TCO superstrates with high haze and broad *ADF* are developed.

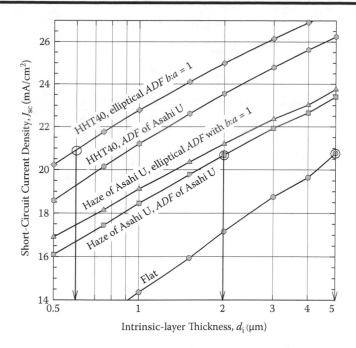

**FIGURE 5.7**    Short-circuit current densities of the analyzed μc-Si:H cell as a function of absorber thickness for different scattering parameters. The arrows denote the thicknesses of the absorber layers of the cells with different scattering properties which render the same level of $J_{SC} = 21$ mA/cm$^2$.

## 5.3    OPTICAL DESIGN OF MULTI-JUNCTION SOLAR CELLS

The concept of a multi-junction solar cell enables a better utilization of the solar spectrum in terms of efficient conversion of a broader part of the spectrum into electrical energy (Green 2005). The gain in conversion efficiency is related to reduction of thermalization losses since more absorber layers with different energy bandgaps are employed, starting with the one with the highest gap ($E_{G1}$) and ending with the one with the lowest gap ($E_{Gn}$), as shown in Figure 5.8.

Such structure ensures that the high energy photons (short-wavelength light) are absorbed in the front absorbers with high energy gap, ensuring a small difference in the photon energy and the energy bandgap of the absorber, which means small thermalization losses. Photons with lower energy (longer-wavelength light) are transmitted to the succeeding absorbers with lower energy bandgaps. A proper distribution of absorbed light has to be ensured in the structure to obtain approximately equal generation of the photocurrent in the multi-junction cell components, which are connected in series. Thus, the photocurrent, presenting $J_{SC}$ in the short-circuit condition, of the entire multi-junction cell is always smaller than that of a single cell concept due to

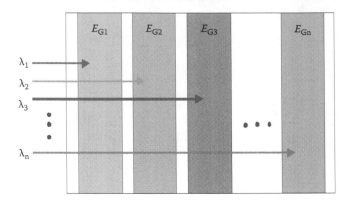

**FIGURE 5.8**   Principle schematic of a multi-junction solar cell with absorbers with different energy bandgaps, $E_{Gi}$, to obtain selective absorption of the solar spectrum ($\lambda_1 < \lambda_2 < \lambda_3 < \lambda_n$).

absorption distribution into more cell components. However, the gain in multi-junction cells is in open-circuit voltage, which is represented by the sum of the voltages corresponding to single cell components connected in series. Thus, lower thermalization losses are finally reflected in higher open-circuit voltages of the devices, with bandgap-engineered absorbers.

In thin-film silicon solar cells, besides single-junction, double-junction (tandem) and triple-junction devices also have reached the production level (Xu et al. 2010; Kluth et al. 2011). Realizations of multi-junction devices in other thin-film technologies (CIGS, CdTe) are also under research and development (Nishiwaki et al. 2003; Schmid et al. 2009; Lipovšek, Krč, and Topič 2010).

Optical modeling and simulations are important tools for designing multi-junction devices. As in the case of single-junction solar cells, we have to ensure efficient light trapping and, consequently, enhanced absorption in absorber layers in the structure. In addition, we have to achieve equal photocurrent generation in each cell component of the multi-junction device (current-matched device), which is linked to a proper determination of thicknesses and energy bandgaps, which relate to the absorption coefficients of the absorbers.

What follows are a few demonstrative examples of application of optical modeling and simulations in thin-film multi-junction solar cells.

### 5.3.1   ABSORBER THICKNESS DETERMINATION IN TANDEM SILICON SOLAR CELLS

As a tandem Si solar cell, a combination of the top cell with an amorphous (a-Si:H) and the bottom cell with a microcrystalline (µc-Si:H) absorber is often used, since the bandgaps of the materials (~1.75 eV for a-Si:H and ~1.2 eV for µc-Si:H) match well with the theoretically determined optimal values of 1.71 eV (top) and 1.14 eV (bottom) for the AM 1.5 solar spectrum.. Solar cells in such

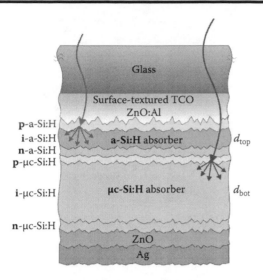

**FIGURE 5.9**   Schematic structure of a micromorph silicon solar cell.

configuration are known as micromorph or hybrid Si solar cells; their basic structure is shown in Figure 5.9.

The top a-Si:H-based cell absorbs mostly the short-wavelength part, whereas the bottom μc-Si:H cell absorbs the remaining long-wavelength part of the spectrum. We can see this also from the $QE$ curves in Figure 5.10.

In the analyzed case the energy bandgaps of the absorbers are defined; thus, only thicknesses of the absorbers are determined in our study to ensure the current matching between the top and the bottom cell. We consider the simulated micromorph solar cell in the following configuration: glass/ZnO:Al TCO (800 nm)/p-a-SiC:H (10 nm)/i-a-Si:H ($d_{top}$)/n-a-Si:H (15 nm)/p-μc-Si:H (10 nm)/i-μc-Si:H ($d_{bot}$)/n-μc-Si:H (15 nm)/ZnO (80 nm)/Ag. Scattering properties of the etched ZnO:Al TCO superstrate with the *rms* surface texture of 80 nm were considered in simulations (Krč et al. 2003). The analyzed cell does not include an intermediate reflector between the top and bottom cells (this is the subject of investigation in the next section). No ARC is applied to the front glass. We carried out optical simulations of the cell with different combinations of $d_{top}$ and $d_{bot}$. $J_{SC}$s of the top and the bottom cell are shown for selected values of $d_{top}$ as a function of $d_{bot}$ in Figure 5.11. Cross-sections between the corresponding $J_{SC\,top}$ and $J_{SC\,bot}$ curves represent the current matching condition. The values of $d_{top}$, $d_{bot}$, and $J_{SC}$ of the current-matched devices are summarized in Table 5.1. The $QE$ results corresponding to the current-matched cells with $d_{top}$ of 150 nm and 250 nm are shown in Figure 5.10. Increasing the thicknesses of the absorber layers, the corresponding $QE$s are raised and the corresponding $J_{SC}$ is enhanced.

With this example we demonstrate that by means of simulations with a calibrated optical model we can determine the values ensuring current matching of tandem device and save on a series of experiments with thickness variations.

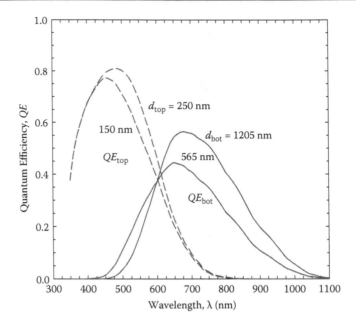

**FIGURE 5.10** The *QE* plots of the top and the bottom cell for two current-matched cases (1 and 3).

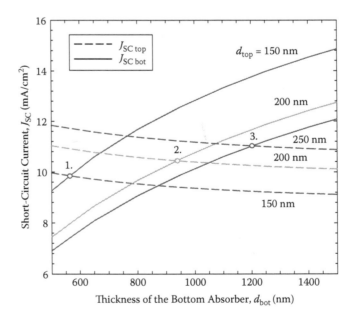

**FIGURE 5.11** Short-circuit current densities of the top and the bottom cell in the micromorph cell as a function of the thickness of the bottom (and top) absorber as the parameter. Cross sections correspond to current-matched devices.

**TABLE 5.1    Specification of Thicknesses for Current-Matched Micromorph Cells Obtained from Figure 5.11**

| Cross-Section No. | $d_{top}$ | $d_{bot}$ | $J_{SC\,top} = J_{SC\,bot}$ |
|---|---|---|---|
| 1. | 150 nm | 565 nm | 9.84 mA/cm² |
| 2. | 200 nm | 940 nm | 10.43 mA/cm² |
| 3. | 250 nm | 1205 nm | 11.03 mA/cm² |

It is good if our simulator has the possibility of an automatic search for current matching. For example, in our simulator Sun*Shine* we implemented a procedure that can find an optimum of certain functions (in this case matching between $J_{SC\,top}$ and $J_{SC\,bot}$) by changing specific input parameters (in this case thicknesses, but it can be absorption coefficients or others). Furthermore, the simulations enable sensitivity analyses of thickness variation (e.g., due to variation of deposition parameters or across larger areas) on performance of tandem structures as an important evaluation step in the validation of production, output power yield, and, nonetheless, cost per watt peak.

## 5.3.2    OPTIMIZATION OF THE INTERMEDIATE REFLECTOR IN TANDEM SILICON SOLAR CELLS

In this section we introduce a single-layer intermediate reflector (interlayer) between the top and the bottom cell in the micromorph cell (Figure 5.12) and

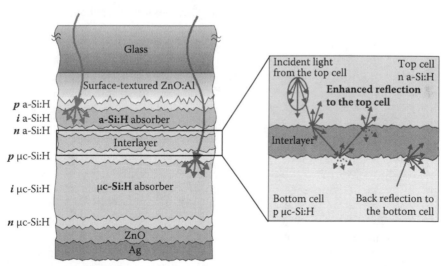

**FIGURE 5.12**   The intermediate reflector (interlayer) in the micromorph silicon solar cell (in this case its texture is a consequence of a textured front ZnO:Al TCO).

investigate its effects. Often a ZnO layer is used as the interlayer (Fischer et al. 1996; Yamamoto et al. 2002). Realizations based on other materials (such as SiO$_x$) have been proposed recently (Buehlmann et al. 2007; Das et al. 2008). The main function of the interlayer is to efficiently reflect short-wavelength light, which was not absorbed within its first pass, back to the top cell. On the other hand, the long-wavelength light has to be transmitted to the bottom cell efficiently. In this way the interlayer enables the use of a thinner i-a-Si:H absorber in the top cell, resulting in a less pronounced Staebler-Wronski effect (Staebler and Wronski 1977) of light-induced degradation of a-Si:H material, which presents the main reason for the decrease in efficiency of the single or tandem cell from its initial to a lower stabilized value. In micromorph cells with an interlayer, therefore, a thinner i-a-Si:H absorber layer ($d < 250$ nm) can be used, leading to a higher stabilized conversion efficiency close to the initial one (Meier et al. 2004).

Besides increased reflection back to the top cell, additional scattering of light can be introduced at the top or the bottom interface of the interlayer if it is textured. In particular, a new type of texturization can be introduced there, tuned to the scattering in the top or bottom cell. One example is an LP-CVD ZnO asymmetric interlayer, where the crystalline growth of the layer introduces new texture in the cell (Söderström et al. 2009).

In the following, we present the results of optical simulations that were employed to optimize optical properties, in particular refractive index, of the interlayer, in order to improve reflection of short-wavelength light back to the top cell.

Figure 5.13 shows refractive indexes of the interlayers, $n_{\text{inter}}$, used in our simulation study. To cover a broad range of values of $n_{\text{inter}}$, we introduced hypothetical interlayers in addition to a magnetron sputtered undoped ZnO layer, which will serve as a reference interlayer. Wavelength-independent behavior was applied to the hypothetical interlayers due to simplicity. In Figure 5.13 the refractive indexes of the surrounding layers of the interlayer in micromorph solar cells, that is, n-a-Si:H and p-µc-Si:H, are plotted as well. For all materials, the values of $n$ that correspond to $\lambda = 650$ nm are given in the figure. Besides refractive index, optical losses were applied to the interlayers by considering extinction coefficients of selected materials as evident from the names of the hypothetical interlayers. For example, the name n2.5kZnO stands for a layer with $n_{\text{inter}} = 2.5$ and extinction coefficient $k_{\text{inter}}$ of ZnO material. As possible interlayers with low refractive indexes ($n_{\text{inter}} < 2$), we selected SiO$_2$ ($n_{\text{inter}} = 1.5$) and MgF$_2$ ($n_{\text{inter}} = 1.4$) material. As already mentioned, SiO$_x$-based materials have been proven in realistic devices as promising low refractive index candidates ($n_{inter} = 1.5$–2) (Buehlmann et al. 2007; Dominé et al. 2008). When selecting the materials for the interlayer, we should be aware that besides required optical properties, sufficient electrical conductivity in a vertical direction of the layer has to be ensured. In some cases this can be achieved by doping of the intrinsic materials. More complicated contacting schemes (e.g., through holes, local contacting) could be used to establish conductive paths through the layer.

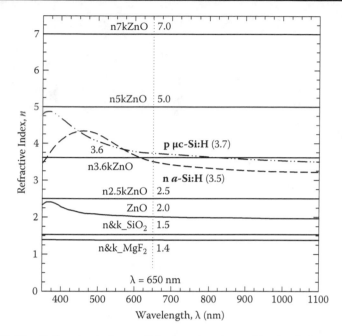

**FIGURE 5.13**    Refractive indexes of the interlayers used in simulations.

We analyze the role of the interlayers in a micromorph solar cell in a similar configuration as in the previous section: glass/ZnO:Al TCO (1 µm)/p-a-Si:H (10 nm)/i-a-Si:H (150 nm)/n-a-Si:H (15 nm)/interlayer/p-µc-Si:H (10 nm)/i-µc-Si:H (2.6 µm)/n-µc-Si:H (15 nm)/ZnO (80 nm)/Ag. The thickness of the interlayer was chosen to be 100 nm. Simulations indicated that since a substantial part of light is scattered in the analyzed solar cell structure, the interference effects related to the interlayer thickness do not take place here. Experimentally determined complex refractive indexes of all layers were used in the simulations (Springer, Poruba, and Vaneček 2004). A highly textured front ZnO:Al TCO with *rms* surface roughness of 100 nm is applied to efficiently scatter the incident light in the structure. Realistic scattering parameters of the ZnO:Al superstrate were used (Krč et al. 2003). The effect of decreased reflectance due to texturization of the thin interlayer was not included in these simulations.

In Figure 5.14 we can see the simulation results $QE_{top}$ and $QE_{bot}$ of the cell without (w/o) and with the ZnO interlayer. In $QE_{top}$ we observe an enhancement in the long-wavelength part ($\lambda = 500$–700 nm, effect a) and consequently we also see a drop in the short-wavelength part of $QE_{bot}$ (effect b) for the cell with the interlayer. Both are related to increased reflectance in the middle of the solar cell structure as a result of the introduced interlayer (as depicted in Figure 5.12).

We investigate the effects of different interlayers whose refractive indexes are shown in Figure 5.13. In Figure 5.15 $QE$ results are shown for the case of hypothetical interlayers with the lowest (n&k_MgF2) and the highest (n7kZnO) refractive index and compared to the cell with the ZnO interlayer. Beneficial

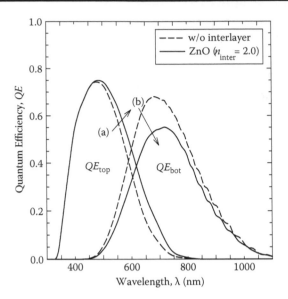

**FIGURE 5.14**    The effects of ZnO interlayer on the *QE* of the top and the bottom cell.

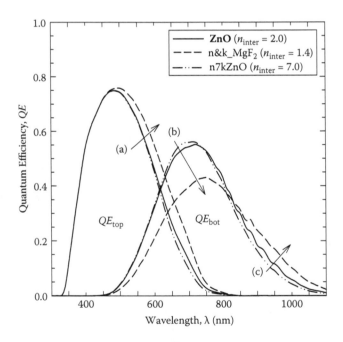

**FIGURE 5.15**    The effects of the interlayers with a high and small refractive index on the *QE* of the top and the bottom cell.

enhancement in the long-wavelength $QE_{top}$ are observed for the n&k_MgF$_2$ interlayer (effect a). No increases in $QE_{top}$ can be observed if applying the n7kZnO interlayer with extremely high $n_{inter}$. By using an interlayer with lower $n_{inter}$ than the n-a-Si:H layer, the effect of total reflection ($R$ close to unity) can occur at the n-a-Si:H/interlayer interface for the scattered light beams approaching under sufficiently large incident angles. This explains why the low refractive index interlayer behaves better than does the high refractive index interlayer in solar cells with textured interfaces.

In addition to the increase in long-wavelength $QE_{top}$ and consequent decrease in short-wavelength $QE_{bot}$ (effect b), we can observe the effect c—an increase in long-wavelength $QE_{bot}$ ($\lambda > 800$ nm). Results of the optical situation confirm that this effect is related to total reflectance $R_{back}$ at the bottom interface of the interlayer (interlayer/p-$\mu$c-Si:H) for the scattered light propagating back from the bottom cell (see Figure 5.12, right side). Similar as in the top cell, in the case of an interlayer with low $n_{inter}$, enhanced confinement of long-wavelength light can be achieved in the bottom cell. This effect can partially compensate the decrease in the shorter wavelength $QE_{bot}$ in the final $J_{SC\,bot}$.

We studied the role of different interlayers on $R_{back}$ of long-wavelength light ($\lambda > 800$ nm) as a function of the incident angle with respect to the normal interface. The results corresponding to $\lambda = 950$ nm are shown in Figure 5.16. For

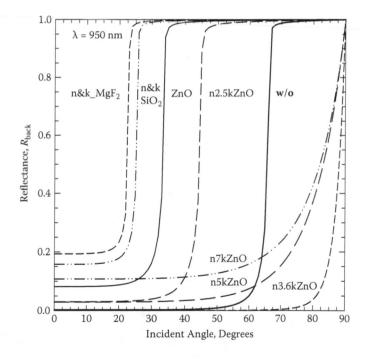

**FIGURE 5.16** Angular dependency of reflectance at different interlayers for the p-$\mu$c-Si:H incident medium.

comparison the $R_{\text{back}}$ of the system w/o interlayer (i.e., p-μc-Si:H/n-a-Si:H interface) is plotted. For the interlayers with $n < n_{\text{p-μc-Si:H}} = 3.6$ at $\lambda = 950$ nm, total reflection ($R_{\text{back}} \rightarrow 1$) is observed before the incident angle of 90 degrees. The smaller the value of $n_{\text{inter}}$ is, the smaller is the angle of total reflection. In this case also scattered beams with a narrower *ADF* characteristic are reflected efficiently.

The trends related to interlayers with different $n_{\text{inter}}$, which we observed for $QE_{\text{top}}$ and $QE_{\text{bot}}$, are reflected in $J_{\text{SC top}}$ and $J_{\text{SC bot}}$. The results are presented in Figure 5.17. The minimum value of $J_{\text{SC top}}$ is detected for the interlayer n3.6kZnO (value of $n_{\text{inter}}$ close to the ones of adjacent silicon layers). A significant increase in $J_{\text{SC top}}$ is observed (up to 25%, ref. cell w/o interlayer) for interlayers with $n < 3.6$ (like SiO$_2$ and MgF$_2$), whereas the increase for interlayers with $n > 3.6$ is smaller. For the $J_{\text{SC bot}}$ we can observe an opposite trend (Figure 5.17), despite effect c. Thus, it is obvious that in selection of a suitable interlayer we have to consider the trade-off between improvements in $J_{\text{SC top}}$ and reduction in $J_{\text{SC bot}}$.

Besides single interlayer, concepts of multilayer intermediate reflectors, which exhibit special wavelength-selective characteristics, are investigated (see possible solution in Section 5.5.2).

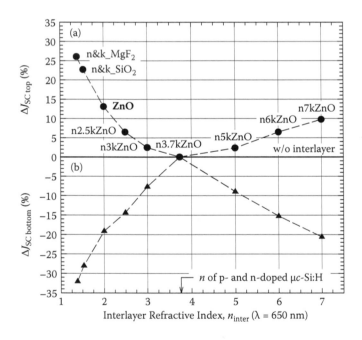

**FIGURE 5.17**   Relative increase/decrease of short-circuit current density in top/bottom cells as a consequence of the introduction of different interlayers in the micromorph solar cell.

### 5.3.3 SIMULATIONS OF NOVEL CONCEPTS OF TANDEM CHALCOPYRITE SOLAR CELLS

Besides silicon-based tandems, other types of materials are of interest for multi-junction solar cells. For example, low-cost polycrystalline chalcopyrite materials such as $Cu(In_x,Ga_{1-x})Se_2$ – CIGS have been recognized for the potential to easily alter their bandgap, enabling better solar spectrum utilization in solar cell applications. Steps toward chalcopyrite tandems are the topic of this section.

By changing the ratio of In/Ga concentrations in chalcopyrite material, bandgaps from 1.02 eV for $CuInSe_2$ (CIS) up to 1.68 eV for $CuGaSe_2$ (CGS) can be achieved (Dullweber et al. 2000). This bandgap alteration ability makes them perfect candidates not only for thin-film single-junction devices but also for tandem and other advanced optical concepts of solar cells.

We analyze here the potential of monolithically stacked CGS top cells and CIGS bottom cells in a tandem configuration from the optical point of view, using 1-D simulations. The effects of the top CGS thickness and the bottom CIGS bandgap are analyzed, and their optimal values for the highest conversion efficiency are determined. More details can be found in Lipovšek (2012).

The Sun*Shine* simulator was employed to determine the $QE$ and $J_{SC}$ of the investigated solar cells. For this determination, a detailed optical analysis combined with a simplified (but justified) electrical analysis is used as in previous sections on silicon cells. Here, as the first step, the external $QE$ is equalized to the calculated absorptance in the CGS or CIGS layer. Then, we include the effects of the non-ideal charge carrier extraction that are present especially in the CGS absorber, taking into consideration the $QE$ data of representative state-of-the-art CGS and CIGS cells (Repins et al. 2008). The $J_{SC}$ is calculated from such determined $QE$ by applying the AM 1.5 solar spectrum. In the simulations, realistic complex refractive indices of the layers—including the sub-bandgap absorption in CGS and CIGS layers (Schmid et al. 2009) and the free carrier absorption in TCO layers—are taken into account.

The conversion efficiencies (*Eff*) of the investigated structures are determined from the simulated $J_{SC}$ by applying the standard one-diode model and then calculating the open-circuit voltage ($V_{OC}$) and the fill factor (*FF*) from the *J-V* characteristics. For these calculations, realistic electrical parameters of the chalcopyrite absorbers [i.e., the bandgap ($E_G$), the saturation current density ($J_0$), and the diode ideality factor (*A*)] are assumed: (i) CGS: $E_G$ = 1.68 eV, $J_0$ = 7·$10^{-7}$ mA/cm², *A* = 2.1 (Young et al. 2003); and (ii) CIGS: $E_G$ = 1.18 eV, $J_0$ = 2.1·$10^{-9}$ mA/cm², *A* = 1.14 (Repins et al. 2008). The saturation current density of CIGS absorbers with bandgaps other than 1.18 eV is extrapolated according to Equation (5.2) (Schmid et al. 2009).

$$J_0 = J_{00} \cdot \exp\left(-E_G/(A \cdot k \cdot T)\right) \qquad (5.2)$$

We investigate the situation and potential of the monolithically stacked CGS/CIGS tandem structure (Figure 5.18).

In a monolithically stacked CGS/CIGS tandem, the top CGS and the bottom CIGS cells are connected optically and electrically with a SnO$_2$:F TCO layer in the presented approach. A thin Mo layer between the CGS and the SnO$_2$:F for improving the ohmic contact of the interface was not considered in the simulations. The transparent front contact is realized as a combination of the intrinsic ZnO and *n*-doped ZnO:Al TCO. To reduce the reflection losses, a thin layer of MgF$_2$ ARC is applied on top of the ZnO:Al TCO layer. The thicknesses of the individual layers are given in Figure 5.18.

We set the bandgap of the top CGS absorber to 1.68 eV, which is close to the theoretically optimal value for the tandems. We varied and optimized two parameters in the analysis: (i) the thickness of the top CGS absorber ($d_{CGS}$), and (ii) the bandgap of the bottom CIGS absorber ($E_{G, CIGS}$). The effects of the $d_{CGS}$ on the *QE* are presented in Figure 5.19 for two selected thicknesses and for $E_{G, CIGS}$ = 1.15 eV. Another effect is pointed out in Figure 5.19: sub-bandgap absorption in the CGS absorber. This parasitic absorption limits the transparency of long-wavelength light ($\lambda$ > 740 nm) to the bottom CIGS absorber and does not contribute to the *QE* of the top cell. It presents the source of additional optical losses that have to be taken into account in the design of chalcopyrite tandems. Increasing the thickness of the CGS absorber results in higher $J_{SC\ CGS}$, but on the other hand parasitic losses are enhanced, reducing the $J_{SC\ CIGS}$.

Let us focus further on the two examples given in Figure 5.19. Based on *QEs* presented for the CGS absorber with the thickness of 1.7 µm, we can determine

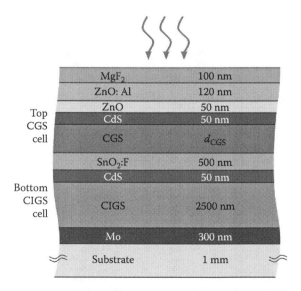

**FIGURE 5.18**    The analyzed concept of a monolithically stacked chalcopyrite tandem.

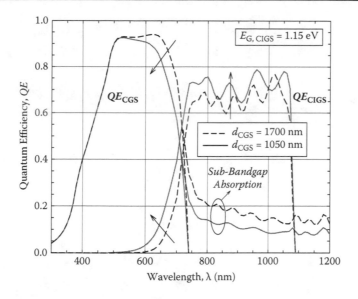

**FIGURE 5.19** The simulated *QE* of CGS/CIGS tandems with different top CGS absorber thicknesses. The effects of thinning down the CGS absorber are indicated by arrows.

the $J_{SC, CGS}$ = 17.5 mA/cm² (AM 1.5) for the top cell. However, due to a relatively thick CGS absorber, the pronounced parasitic sub-bandgap absorption of this material severely hinders the transparency of the top cell in the long-wavelength region. Thus, the $J_{SC}$ of the bottom CIGS cell is limited to $J_{SC, CIGS}$ = 14.1 mA/cm². The simulation results show that in this case, 5.1 mA/cm² of the potential $J_{SC}$ is lost because of the sub-bandgap absorption in the top CGS absorber. This seems to be the problem here. How do we solve it? To minimize optical losses and to ensure better current matching, the $d_{CGS}$ needs to be optimized (decreased). The optimization results indicate that if we reduce $d_{CGS}$ to 1050 nm, we obtain a lower $J_{SC, CGS}$ = 16.4 mA/cm² but also less pronounced sub-bandgap absorption. In this case, only 3.3 mA/cm² of the potential $J_{SC}$ is lost because of the sub-bandgap absorption in the top CGS cell. Therefore, the $J_{SC, CIGS}$ is increased to 16.3 mA/cm². Comparing the $J_{SC, CGS}$ and $J_{SC, CIGS}$, it can be observed that now the top and the bottom cell are current-matched, which is, as mentioned in the previous section, of primary importance for efficient tandems. Thus, the thickness $d_{CGS}$ of 1050 nm was indicated as the optimal value for our tandem.

The bandgap of the bottom CIGS absorber, on the other hand, affects both the $J_{SC}$ and the $V_{OC}$ of the cell, according to the diode model. Higher $E_{G, CIGS}$ results in a higher $V_{OC}$ but lower $J_{SC, CIGS}$. Therefore, because both the $d_{CGS}$ and the $E_{G, CIGS}$ affect the performance of the tandem, they need to be optimized simultaneously, as presented in the following.

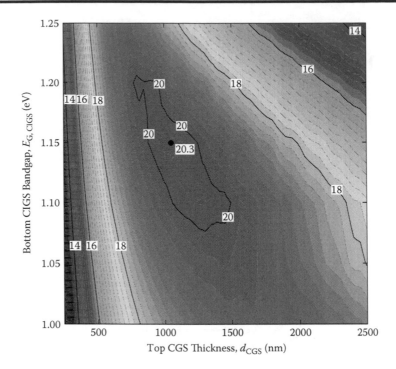

**FIGURE 5.20**   The calculated conversion efficiencies of CGS/CIGS tandems for different $d_{CGS}/E_{G, CIGS}$ combinations. The maximal efficiency of 20.3% can be achieved, considering the realistic state-of-the-art parameters for CGS and CIGS cell.

To determine the optimal values of the $d_{CGS}$ and $E_{G, CIGS}$, rendering the highest *Eff* of the tandem cell, we carried out an extensive set of simulations in which parameters were varied from 100 to 2500 nm and from 1.00 to 1.25 eV, respectively. The results presented in Figure 5.20 indicate that by using the realistic optical and electrical parameters of the CGS and the CIGS cell, the maximal tandem efficiency of 20.3% can be achieved. Furthermore, we also observe that the area with high efficiencies (*Eff* > 20%) is localized tightly around the optimal values, which are $d_{CGS} \approx 1050$ nm and $E_{G, CIGS} \approx 1.15$ eV.

To indicate possible further improvements in the tandem performance, we assumed an optically and electrically idealized CGS cell in the CGS/CIGS tandem in our simulations. In particular we assume two cases of idealization. First, we excluded the sub-bandgap absorption in the CGS absorber and second, we apply the electrical parameters of the state-of-the-art CIGS cell to the CGS cell. Simulations showed that in the case of the idealized CGS cell, a much higher conversion efficiency of the tandem (*Eff* = 27.5%) could be achieved. Thus, improving the electrical and optical performance of the top CGS cell presents a crucial step for highly efficient chalcopyrite tandems. This statement can be generalized also to other types of tandem cells. Upgrades of single-junction solar cells into tandem configurations are worth doing if both cell components exhibit sufficient optical and electrical properties. For example, having a highly efficient cell as the bottom cell and less efficient cell as the top cell, we should be aware that the top cell can absorb the short-wavelength part of the spectrum that may be more efficiently transferred into electrical energy in the bottom cell (despite thermalization losses). If the top cell has poor electrical properties, it should be at least highly transparent so that the light can pass to the bottom cell. The trade-off in the thermalization losses in the bottom cell and losses (optical and electrical) in the top cell has to be found, ensuring higher conversion efficiency of the tandem, compared to the efficiencies of a single cell.

In case of chalcopyrite tandems, further advanced optical solutions as well as improvements of electrical properties of the top cell are needed. Some possible solutions based on the introduction of metal nano-particles (Schmid et al. 2011) and wavelength-selective intermediate reflectors (Lipovšek, Krč, and Topič 2010) have been investigated. A possible alternative for the top cell could also be a CdTe-based solar cell.

### 5.3.4   EFFECTS OF OPTICAL AND ELECTRICAL IMPROVEMENTS IN TRIPLE-JUNCTION SILICON SOLAR CELLS

To show the applicability of 1-D modeling and simulation to triple-junction solar cells, we analyzed a silicon-based cell with a wide-bandgap top a-Si(C):H, a-SiGe:H middle, and μc-Si:H bottom cell, which is also a candidate for mass production (Guha and Yang 2010). The difference between the tandem micromorph solar cell in the standard a-Si:H/μc-Si:H configuration and the triple-junction a-Si(C):H/a-SiGe:H/μc(nc)-Si:H solar cell is the middle a-SiGe:H–based cell, which is introduced to increase the open-circuit voltage of the device and, thus, raise the conversion efficiency further. So far, the highest initial conversion efficiencies of the optimized state-of-the-art a-Si:H/a-SiGe:H/μc-Si:H solar cells of more than 15% have been reported (Xu et al. 2010). To indicate the possibilities and to show the directions for increasing the efficiency of the solar cell further, simulations were used.

In this section, in contrast to other sections, we present the results of both optical and detailed electrical simulations. Electrical simulations were carried out with the numerical simulator ASA (Zeman et al. 1997; Pieters, Krč, and Zeman 2006). Besides the presented results on cell performance, this work is also a demonstration of how the two types of simulations can be combined in the analysis.

Because the amorphous absorbers (intrinsic i-a-Si:H and i-a-SiGe:H layers) are relatively thin ($d_i < 300$ nm) in a-Si:H/a-SiGe:H/μc-Si:H solar cells investigated in our simulations, the calculated initial conversion efficiencies can be considered as the stabilized efficiencies (Meier et al. 2004) (because of the less pronounced effect of light-induced degradation in thin amorphous layers).

Realistic optical and electrical input parameters describing the layers and interfaces in the solar cell are used in simulations (Springer, Poruba, and

Vaneček 2004). In the optical optimization of the solar cell, we apply enhanced scattering properties of the interfaces, low absorption coefficients in the supporting (non-active or partially active) layers (ZnO:Al TCO, *p*- and *n*-doped layers), high reflectivity of the back reflector, and ARCs in order to increase the short-circuit current density, $J_{SC}$, of the solar cell. The influence of each optical parameter in the increase of the $J_{SC}$ in the optimized cell is revealed. In electrical optimization of the solar cell, we increase the bandgap of the *p*-doped layer of the top cell and introduce buffer layers with a low defect density at the p/i interfaces of the top and middle cells to increase the open-circuit voltage, $V_{OC}$, and fill factor, *FF*. Applying these steps, a potential to reach an efficiency above 17% for the a-Si:H/a-SiGe:H/μc-Si:H solar cell is demonstrated.

The solar cell in the following configuration was used as a starting point: glass/ZnO:Al (500 nm)/p-a-SiC:H (10 nm)/i-a-Si:H ($d_{top}$)/n-a-Si:H(15 nm)/p-a-SiC:H (10 nm)/i-a-SiGe:H ($d_{mid}$)/n-a-Si:H(15 nm)/p-μc-Si:H (10 nm)/i-μc-Si:H ($d_{bot}$)/n-a-Si:H (15 nm)/ZnO/Ag (shown in Figure 5.21).

The thicknesses of the absorber layers in the starting cell are $d_{top}$ = 180 nm, $d_{mid}$ = 220 nm, and $d_{bot}$ = 2.4 μm, ensuring a current matching between all three p-i-n component cells. In the top cell a wide bandgap i-a-Si:H absorber is used ($E_G$ = 1.9 eV) to absorb short-wavelength light only and to enable efficient passing of the middle- and long-wavelength light to the middle (i-a-SiGe:H) and bottom (i-μc-Si:H) absorbers. The bandgap of the i-a-SiGe:H and μc-Si:H absorbers were set to $E_G$ = 1.5 eV and 1.3 eV, respectively. a-SiGe:H with $E_G$ of 1.5 eV can be fabricated with sufficiently low defect density. Glass coated with surface-textured magnetron sputtered ZnO:Al TCO is used as a solar cell carrier. The *rms* roughness of

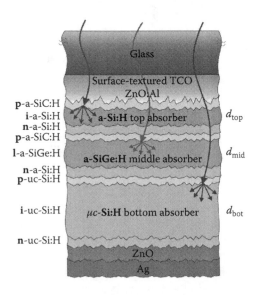

**FIGURE 5.21**    Schematic structure of silicon triple-junction solar cell with a-Si:H, a-SiGe:H, and μc-Si:H absorbers.

80 nm was assumed for the ZnO:Al surface texture and transferred to all subsequent internal interfaces in the solar cell. The ZnO/Ag back reflector (BR) is used at the back side of the cell. No intermediate reflectors (interlayers) are applied between the individual component cells. Realistic complex refractive indexes of the layers (Springer, Poruba, and Vaneček 2004), measured light-scattering parameters of the TCO superstrate (Krč et al. 2003), and realistic reflectance of the textured ZnO/Ag back reflector (Springer et al. 2004) are used in the simulations.

In Figure 5.22 total reflectance, $R_{tot}$, optical losses in the non-active layers (TCO superstrate, BR) and partially active supporting layers ($p$- and $n$-doped layers), and absorptances in the absorber layers, $A_{i,top}$, $A_{i,mid}$, and $A_{i,bot}$, of the starting solar cell are presented. By assuming an ideal extraction of generated charge carriers from the absorber layers and neglecting contributions from the $p$- and $n$-doped layers, $A_{i,top}$, $A_{i,mid}$, and $A_{i,bot}$ can be identified as the external quantum efficiencies $QE_{top}$, $QE_{mid}$, and $QE_{bot}$ of the solar cell (Zeman et al. 2000). Detailed optical and electrical analysis of the state-of-the-art double-junction micromorph solar cell indicated only small differences between absorptances and quantum efficiency curves and, thus, confirmed this assumption.

Optical losses due to reflected light and the light absorption in the non-active and partially active layers can be represented in terms of short-circuit current density losses ($J_{SC,loss}$). A potential short-circuit current density ($J_{SC,potential}$) can

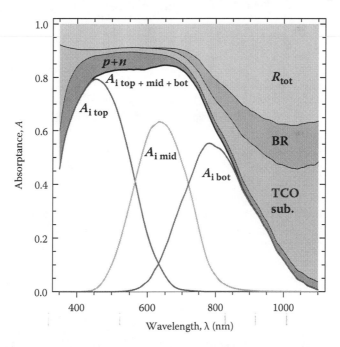

**FIGURE 5.22**  Absorption losses and absorptances in the active (absorber) layers for the starting a-Si:H/a-SiGe:H/μc-Si:H solar cell ($d_{top}$ = 180 nm, $d_{mid}$ = 220 nm, $d_{bot}$ = 2.4 μm).

**TABLE. 5.2    Optical Losses Represented in Short-Circuit Current Density ($J_{SC,loss}$) and a Potential Short-Circuit Current Density ($J_{SC,potential}$) Generated in the Absorber Layers for the Starting and Optically Improved Solar Cells (under STCs)**

| $J_{SC,loss/potential}$ (mA/cm²) | Starting Cell | Optically Improved Cell |
|---|---|---|
| $R_{tot}$ | 7.94 | 4.97 |
| TCO sub. (glass/ZnO:Al) | 4.84 | 1.32 |
| BR (ZnO/Ag) | 2.45 | 1.66 |
| p+n (sum) | 1.95 | 1.05 |
| i-a-Si:H (top absorber) | 8.36 | 11.06 |
| i-a-SiGe:H (middle absorber) | 8.41 | 11.05 |
| i-μc-Si:H (bottom absorber) | 8.47 | 11.01 |

be calculated from $A_{i\,top}$, $A_{i\,mid}$, and $A_{i\,bot}$. The $J_{SC,loss}$ and $J_{SC,potential}$ are calculated for standard AM 1.5 illumination (considering a wavelength range $\lambda = 350–100$ nm) and are presented in Table 5.2. We can see that in the case of the starting solar cell, the highest optical losses are originating from $R_{tot}$ (7.94 mA/cm²), followed by the absorption in the TCO superstrate (mainly in the ZnO:Al layer), textured BR, and *p*- and *n*-doped layers. $J_{SC,potential}$ of ~ 8.4 mA/cm² is obtained for the current-matched starting solar cell.

Optical generation-rate profile in the semiconductor layers was imported into the simulator ASA to carry out the electrical simulations of the cell. In electrical simulations, standard electrical parameters of the amorphous and microcrystalline silicon layers and tunnel recombination junctions between the component cells were used (Zeman et al. 1997). The *J(V)* characteristic of the starting cell, calculated for AM 1.5 illumination, is shown in Figure 5.24 (full black curve). The corresponding external electrical parameters of the cell are given in Table 5.3. Conversion efficiency of 11.4% is calculated for the starting solar cell under standard test conditions (STCs). These values of the external parameters represent the starting point for indicating requirements, leading to better solar cell performance.

To reduce optical losses in the solar cell, the following improvements were assumed in optical simulations:

■ Enhanced scattering parameters to improve light trapping in the cell:
  (i) Ideal haze parameter (scattering level) for reflected and transmitted light, $H = 1$
  (ii) Broad (Lambertian) angular distribution function (*ADF*) of scattered light in reflection and transmission at all textured interfaces

**TABLE 5.3    External Electrical Parameters of the Starting, Optically Improved, and Optically + Electrically Improved a-Si:H/a-SiGe:H/µc-Si:H Solar Cells**

|  | Starting Cell | Optically Improved | Optically + Electrically Improved |
|---|---|---|---|
| $J_{SC}$ [mA/cm$^2$] | 8.32 | 10.82 | 10.88 |
| $V_{OC}$ [V] | 1.95 | 2.01 | 2.21 |
| FF | 0.70 | 0.69 | 0.72 |
| **Eff [%]** | **11.4** | **15.0** | **17.3** |

- Reduced absorptions in non-active layers—significantly decreased absorption coefficients, $\alpha$
  - (iii) Fivefold decrease in $\alpha$ of the front ZnO:Al TCO for all wavelengths
  - (iv) Fivefold decrease in $\alpha$ of the $p$- and $n$-doped a-Si:H and µc-Si:H layers for all wavelengths
- Improved back reflector:
  - (v) BR with an enhanced reflectance of 98%
- Improved light in-coupling:
  - (vi) Optimized single-layer ARC on the top of the glass superstrate with a refractive index of 1.25 and a thickness of 100 nm (Krč, Smole, and Topič 2006) and an antireflective interlayer (such as TiO$_2$) at the ZnO:Al/p-a-SiC:H interface (see details in Section 5.4.2)
  - (vii) Adjustment of the absorber thicknesses to ensure sufficiently high $J_{SC,potential}$ and current matching in the optimized triple-junction solar cell

High $H$ and broad $ADF$ can be achieved by further optimization of the morphology of textured substrates and interfaces. Besides random roughness, substrates with periodical surface texture indicated potential for effective light scattering (Isabella et al. 2008). Further on, TCOs with low free-carrier concentration, resulting in a lower optical absorption in the long-wavelength region, are under investigation (Selvan et al. 2006). Assuming that a significant reduction of absorption coefficients of $p$- and $n$-doped layers may be a critical issue, the use of very thin efficiently doped layers are of interest. High reflectance of the back reflector may be achieved using some specially designed dielectric mirrors (Isabella, Krč, and Zeman 2010), avoiding surface plasmon absorption at textured metals. Glass superstrates covered with a single-layer ARC as well as multilayer broadband ARCs can be used (Meier et al. 2004). Significant improvements in light in-coupling related to internal ARCs (such as TiO$_2$) at TCO/p interfaces have already been reported for the single-junction silicon thin-film solar cell (Fujibayashi, Matsui, and Kondo 2006). The adjustment of

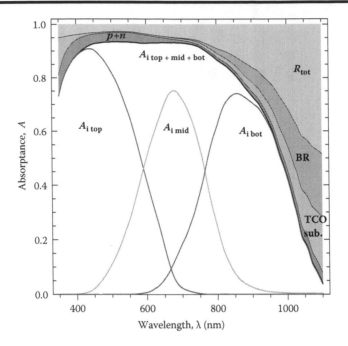

**FIGURE 5.23** Absorption losses and absorptances in the absorber layers for the optically improved a-Si:H/a-SiGe:H/µc-Si:H solar cell ($d_{top}$ = 110 nm, $d_{mid}$ = 280 nm, $d_{bot}$ = 2.1 µm).

the thickness of absorbers for the current matching can be defined and carried out using optical simulations.

The results of optical simulation of the optically improved solar cell, including the improvements (i-vii) are shown in Figure 5.23 and Table 5.2. The adjusted thicknesses in the optimized solar cell are $d_{i\,top}$ = 110 nm, $d_{i\,mid}$ = 280 nm, and $d_{i\,bot}$ = 2.1 µm. We can see that considering the improvements (i-vi), the thickness of the absorbers in the top and bottom cells can be reduced (especially in the top cell), ensuring high $J_{SC,potential}$ ≈ 11 mA/cm² at the same time. However, to achieve sufficiently high absorptance in the i-a-SiGe:H absorber (middle cell), the increase of its thickness from $d_{i\,mid}$ = 220 nm (starting cell) to 280 nm was required. The introduction of an intermediate reflector between the middle and the bottom cell (such as thin ZnO layer) may help to reduce $d_{i\,mid}$ (not investigated here).

Absorption losses ($R_{tot}$, TCO superstrate, BR, $p + n$) in the optically improved solar cell are noticeably reduced (Figure 5.23, Table 5.2), whereas absorptances $A_{i\,top}$, $A_{i\,mid}$, and $A_{i\,bot}$ and corresponding $J_{SC,potential}$ values are consequently improved. To indicate the importance of each optical improvement (i-vii) in the optically improved solar cell, a comparative study was carried out. The results are presented in Figure 5.24. The heights of the bars represent the relative contribution (in %) of each individual improvement to the increase of $J_{SC,potential}$ achieved with the optimized cell. The increase is considered as the difference

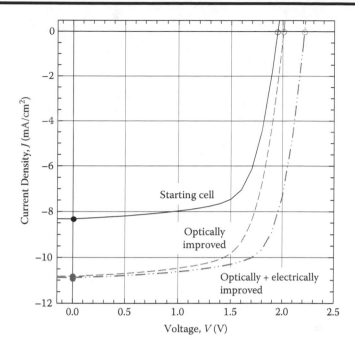

**FIGURE 5.24**  Current-voltage characteristic for the starting, optically improved, and optically and electrically improved a-Si:H/a-SiGe:H/µc-Si:H solar cell (under STCs).

between $J_{SC,potential}$ of the optimized and starting cells ($\Delta J_{SC,potential} \approx 2.6$ mA/cm$^2$ for all component cells). The bars indicate how much of the increase we would lose (relatively) in the given component cell, if a certain improvement was excluded from the others in the optically improved solar cell (see Figure 5.25).

For the top and the bottom cell the most important role is indicated for the broad *ADF* of scattered light. Here the initial elliptical shape of the *ADF* (typical for etched ZnO:Al) (Krč et al. 2003) used in the simulation of the starting cell was broadened to the Lambertian distribution (ellipsis with $b{:}a = 1$). The main contribution to $J_{SC,potential}$ of the middle cell is revealed for increased thickness of the middle a-SiGe:H absorber layer (220–280 nm). By weighting the importance of the optical improvements, we have to mention that there exists a strong synergy between the improvements, as indicated by simulations (e.g., without having a high *H*, the role of the *ADF* is lowered).

The electrical simulations of the optically improved solar cell were carried out using the same electrical input parameters as for the starting solar cell at first (no electrical improvements so far). The results of electrical simulations, *J(V)* characteristic, and the external electrical parameters of the optically improved solar cell are presented in Figure 5.24 (dashed curve) and Table 5.3. A significant increase in the $J_{SC}$ (from 8.32 mA/cm$^2$ to 10.82 mA/cm$^2$) is observed as a consequence of enhanced $A_{i,top}$, $A_{i,mid}$, and $A_{i,bot}$. Due to increased $J_{SC}$, the

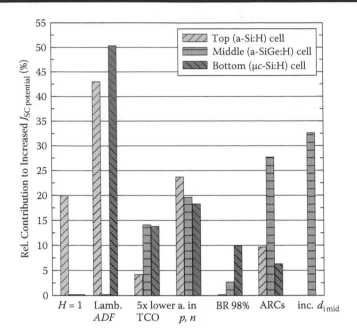

**FIGURE 5.25**   Relative contribution of each optical improvement to $J_{SC,potential}$ of the component cell in the optically improved a-Si:H/a-SiGe:H/μc-Si:H solar cell.

efficiency of the optically improved solar cell is improved noticeably (from 11.4% to 15.0%). The $V_{OC}$ and *FF* of the solar cell are changed slightly.

To raise the efficiency of the solar cell further, besides optical also electrical improvements applied to the cell were investigated. We introduced two electrical improvements in the simulations:

(i) The bandgap of the *p*-doped a-SiC:H layer of the top cell was increased from 1.9 eV (starting cell) to 2.2 eV.

(ii) Thin (5nm) i-a-Si:H buffer layers with a large bandgap (1.9 eV top cell, 1.8 eV middle cell) but low defect density were introduced at the p/i interfaces of the top and the middle cell.

The rest of the electrical material parameters we kept unchanged (values as in the case of the starting solar cell). The simulated *J(V)* characteristic and external electrical parameters of the optically and electrically improved solar cell are given in Figure 5.23 (dash-dot-dot curve) and Table 5.3. The electrical improvements result in an increased $V_{OC}$ (from 2.01 V to 2.21 V) and in improved *FF* (from 0.69 to 0.72). From these values the efficiency of 17.3% is calculated for the optimized cell under STCs. It is important to note that in the electrical simulations, the input parameters referring to average quality of layers were considered (not state-of-the-art materials with advanced properties).

By improving the quality of materials (e.g., lowering defect density, increasing electron and hole mobility), there is still space for further electrical improvements. For example, the simulations indicated that by decreasing the concentration of dangling bonds in the bottom μc-Si:H absorber (from $4 \times 10^{15}$ cm$^{-3}$ to $1 \times 10^{15}$ cm$^{-3}$) the *FF* of the optimized cell can be improved significantly (up to 0.80). Thus, if the optical requirements (improvements i-vii) cannot be reached, the use of high-quality a-Si:H and μc-Si:H films can contribute to additional electrical improvements. In this way the efficiency of 17% of the optimized a-Si:H/a-SiGe:H/μc-Si:H solar cells can be exceeded.

## 5.4 THE ROLE OF ANTIREFLECTIVE COATINGS IN THIN-FILM SILICON SOLAR CELLS

### 5.4.1 THE PRINCIPLE OF ANTIREFLECTIVE COATINGS

Antireflective coatings and structures (textures) are applied to top interfaces of solar cells to decrease reflectance of incident light and thus enable more light to enter the solar cell structure. In this way absorption of light in the absorber layers can be increased. There are two main principle solutions to decrease the reflection:

■ coatings (single- or multilayer) and
■ textures (nano and geometrical optics).

In this section we focus on the first solution. (We address the second solution in Chapter 6.) In both solutions, except in the case of geometrical optics, the same principle is followed to obtain smooth gradual transition from the refractive index of incident medium to the one of the medium in transmission. In the case of large textures where geometrical optics takes place, the reflected beam usually has the opportunity to enter the structure at the next surface corrugation once again.

For ARCs, simulations are of great support to the analytical methods; they are useful especially in the design of multilayer coatings. Let us explain the principle of the antireflection effect on the example of a single-layer coating as we see in Figure 5.26.

The incoming light wave experiences reflection at the front and back interface of the ARC. The initial reflectance (interface medium 1/medium 2) can be decreased in this structure with the ARC, if the reflected waves from both interfaces of the ARC have opposite phases, enabling suppression or full cancellation of the waves propagating back in medium 1. If the phase shift is exactly 180 degrees and if the amplitudes of the shifted waves are equal, zero reflectance can be achieved. This condition can be obtained if the thickness and refractive index of the ARC are determined as specified in Equation (5.3).

$$d_{\text{ARC opt}} = \frac{\lambda_0}{4 n_{\text{ARC opt}}}, \qquad n_{\text{ARC opt}} = \sqrt{n_1 \cdot n_2}, \qquad (5.3)$$

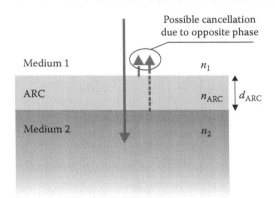

**FIGURE 5.26**   Principle of a single-layer antireflective coating.

We should be aware that the optimal condition of zero reflectance can be achieved for a single wavelength. However, significant suppression in the reflectance is present also in a broader wavelength range. Furthermore, weakly absorbing materials have to be used for ARCs to avoid optical losses therein. Thus, their optical properties can be represented by the real part ($n$) of the complex refractive index ($N$) only, as considered in Equation (5.3).

To demonstrate the ARC effect, we take a μc-Si:H (absorber) layer as medium 2 and air as medium 1. In simulations, the thickness of the μc-Si:H layer was considered to be sufficiently large to avoid reflections from the back interface of the layer. The simulated reflectance of the front air/μc-Si:H interface (structure w/o ARC) is presented in Figure 5.27 by a full line. We can observe a relatively high level of reflectance (>0.3) as a consequence of a relatively large difference in refractive indexes of air ($n = 1$) and μc-Si:H material (n ≈ 4 at λ = 550 nm). Next, we introduce a single-layer ARC on top of the layer. Its thickness and refractive index are determined according to Equation (5.3) by considering minimal reflectance at the wavelength of λ = 550 nm. This wavelength was selected as a representative of the wavelength region in which the solar spectrum has the highest intensity.

The optimal single-layer ARC for the analyzed structure should have $n_{ARC}$ ≈ 2 and d ≈ 70 nm. The refractive index corresponds to $Si_3N_4$ or any other material with similar refractive index and low optical losses (ZnO is also a possible candidate). As shown in Figure 5.27, almost zero reflectance can be obtained around λ = 550 nm for the $Si_3N_4$-based ARC. Reflectance is significantly decreased in the entire scope of wavelengths.

Our aim is to make the region of very low reflectance even broader. This can be achieved by introducing multilayer ARCs. Here, more reflected waves are present in the incident medium, since the number of interfaces in the ARC increases. The situation with cancellation of reflected rays becomes more complex, but we can approach the optimal condition in a broader range of

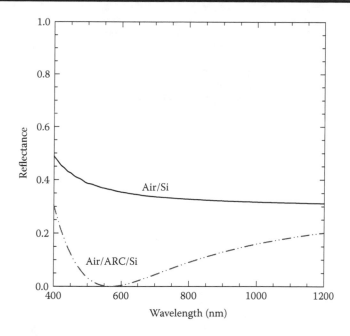

**FIGURE 5.27**    Simulated reflectance of a flat μc-Si:H layer and the effect of a single layer of $Si_3N_4$ ARC ($n_{ARC} \approx 2$, $d_{ARC} = 70$ nm).

wavelengths. A simulation example is given for the case of a double-layer ARC on glass superstrate in this case. The simulated reflectance of the air/ARC1/ARC2/glass stack, where both ARCs were optimized by means of optical simulations, is presented in Figure 5.28 and compared to the reflectance of the air/glass and air/ARC/glass optical system. The specifications of optimal ARCs are given in the figure caption. Low reflectance can be achieved in a much broader wavelength region than with a single-layer ARC. With multilayer ARCs (more than two ARCs in a stack), we can obtain even better results.

## 5.4.2   THE EFFECT OF ANTIREFLECTIVE COATINGS IN THIN-FILM SOLAR CELLS

In the case of thin-film solar cells, light does not enter the semiconductor layers directly from air but through the front glass (superstrate of the cell or encapsulation glass in PV modules) and TCO contact. From the optical point of view, glass with $n \approx 1.5$ and TCO with $n \approx 2$ already act as antireflective layers, since their refractive indexes are between that of air (1.0) and of semiconductor. However, their thicknesses are not tuned to the antireflection condition. Improvements in light in-coupling by introducing ARCs at front interfaces are still important to enhance the performance of the solar cells.

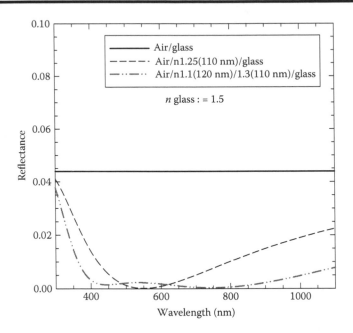

**FIGURE 5.28** The effect of a single- ($n_{ARC}$ = 1.25, $d_{ARC}$ = 110 nm) and a double-layer ARC ($n_{ARC1}$ = 1.1, $d_{ARC1}$ = 120 nm, $n_{ARC2}$ = 1.3 and $d_{ARC2}$ = 110 nm) on the reflectance on glass substrate.

Next, we show three simulation examples related to ARCs in thin-film solar cells. In the first example we demonstrate how the front glass (in combination with EVA foil) contributes to better light in-coupling in CIGS solar cells. The second example deals with optimization of ARC for the tandem micromorph cell. In the third example the effect of an introduced $TiO_2$ layer between $SnO_2$:F TCO and p-a-SiC:H layer on the performance of an a-Si:H solar cell is shown.

The basic structure of the first simulation case is as follows: ZnO:Al (400 nm)/ZnO (50 nm)/CdS (40 nm)/CIGS (1.8 µm)/Mo/substrate. CIGS solar cells are usually realized in a substrate configuration; therefore, glass superstrate is not present on the top of the laboratory cells. However, in PV modules, EVA foil and encapsulation glass are added on the top of the ZnO:Al TCO. If not even, some additional encapsulation layers are added as moisture-blocking layers.

In Figure 5.29 we show the simulation of the CIGS solar cell without (lab cell) and with glass/EVA encapsulation (PV module). In this simulation we used the same refractive index for glass and EVA foil ($n$ = 1.53).

Improvements in *QE* ($\lambda$ = 500–800 nm) related to the antireflection effect of the glass/EVA stack can be observed for the encapsulated module, leading to enhancement in $J_{SC}$ of ~ 2%. Encapsulation is not a perfect antireflection solution, which is demonstrated by simulation of a case where an $MgF_2$ layer

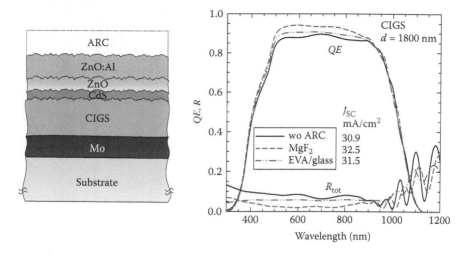

**FIGURE 5.29**    Schematic of the solar cell structure (left) and quantum efficiency and total reflectance of a CIGS solar cell with different antireflective configurations at the front side of the structure (right).

with a thickness of 70 nm is placed on the top of the ZnO contact of the lab cell. The MgF$_2$ coating has a refractive index of 1.4, which is closer to the optimal value determined for the air/ZnO ($n_{ZnO} \approx 2$) interface [$n_{ARC\,opt} = 1.41$, according to Equation (5.3)] and its thickness enables the reflectance minimum at 550 nm.

Further on, reflectance of the air/glass interface can be minimized by introducing ARCs on the glass surface. By considering the refractive index of glass $n = 1.53$, the reflectance of air/glass interface is 0.044 (4.4%). We investigate the effects of minimization of this reflectance on solar cell performance in the following. A tandem micromorph solar cell in the following configuration is analyzed: (ARC)/glass/ZnO:Al (1 µm)/p-a-SiC:H (10 nm)/i-a-Si:H (150 nm)/n-a-Si:H (15 nm)/ZnO (100 nm)/p-µc-Si:H (10)/i-µc-Si:H (3 µm)/n-µc-Si:H (15 nm)/ZnO (80 nm)/Ag. Realistic optical properties of the layers and scattering parameters of textured interfaces ($\sigma_{rms}$ of the TCO superstrate was 80 nm) were considered in simulations (Krč et al. 2003; Springer, Poruba, and Vaneček 2004). We study the effects of a single-layer ARC on glass on the QE and $J_{SC}$ of top and bottom cells. In simulations we varied the refractive index and thickness of the coating systematically, as shown in Figure 5.30 (points on the grid present combinations of $n_{ARC}$ and $d_{ARC}$ that we used in simulations).

Here we performed simulations for all 48 combinations since the simulator Sun*Shine* that was employed in the study facilitates automatic iterative execution of simulations for a predefined grid of parameters. Moreover, automatic optimization of parameters is possible where fewer simulations can be executed

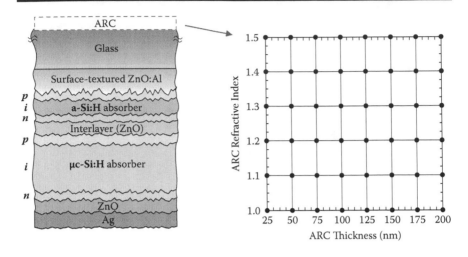

**FIGURE 5.30**  Antireflective coating on glass superstrate in micromorph solar cell and combinations of the refractive index and thickness of the single-layer ARC used in the presented simulation study.

to find the optimal combination (not used here). For the analyzed micromorph solar cell without any ARC on top of the glass, the simulated values of short-circuit current densities were $J_{SC\,top} = J_{SC\,bot} = 10.1$ mA/cm$^2$ (current-matched device). In the case of tandem devices we can optimize the ARC to obtain the highest improvement either in the performance of the top or the bottom cell, depending on our requirements for a particular cell. We can achieve this by tuning the thickness and refractive index of the ARC. For the top cell the minimum of the reflectance we define at $\lambda \approx 500$ nm, whereas for the bottom cell we define it at $\lambda \approx 750$ nm. The results of simulated reflectance of the air/glass and optimized air/ARC/glass optical system and the corresponding improvements in $QE_{top}$ and $QE_{bot}$ of the cells are shown in Figure 5.31.

In Table 5.4 the optimal combinations of $n_{ARC}$ and $d_{ARC}$ and the corresponding $J_{SC}$ increases are specified. In both cases for the ARC, $n_{ARC} \approx 1.2$ is found to be an optimal value, whereas the difference in optimal thickness (100 nm and 150 nm) causes the required shift in the minimum of reflectance.

So far we have investigated an ARC on top of the device. However, internal interfaces that are located in front of the absorber layer also cause reflection losses. As the third simulation case, we show the effects related to the introduction of a single-layer ARC at an internal interface of a solar cell—in particular, a TiO$_2$ ARC layer between the SnO$_2$:F TCO and p-a-SiC:H layer in a single-junction a-Si:H cell. In simulations we assume $\sigma_{rms}$ surface texture of the TCO of 40 nm. We simulated the solar cell in configuration glass/SnO$_2$:F (650 nm)/TiO$_2$ ARC (5 nm)/p-a-SiC:H (10 nm)/i-a-Si:H (300 nm)/n-a-Si:H (15 nm)/ZnO (80 nm)/Ag. Scattering parameters of the Asahi U-type TCO were considered in simulations (Krč et al. 2002). The texturization itself affects the antireflective

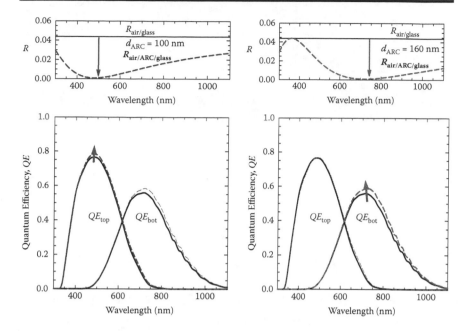

**FIGURE 5.31**   The effects of a single-layer ARC on glass optimized for top cell (left) and bottom cell (right) on reflectance of air/(ARC)/glass system (top graphs) and on QEs (bottom graphs).

behavior of the interfaces; however, the benefits of the $TiO_2$ ARC remain evident. For $TiO_2$ we assume the refractive index of 2.4 ($\lambda$ = 550 nm), which is between the $n$ of $SnO_2$:F ($n$ = 1.9 at $\lambda$ = 550 nm) and that of p-a-SiC:H ($n$ = 3.9 at $\lambda$ = 550 nm). Therefore, the antireflection effect can be achieved. As already proven in the fabricated cells (Fujibayashi, Matsui, and Kondo 2006), an introduced $TiO_2$ layer also does not worsen the electrical properties of the TCO/p-a-SiC:H if the layer is thin enough. The results of simulated QE are shown in Figure 5.32. Improvements related to the introduced $TiO_2$ are present, especially in the short-wavelength part of the QE.

**TABLE 5.4    Relative Increases in Short-Circuit Current Density of the Top and Bottom Cells for the ARC, Optimized for the Top and Bottom Cells**

|  | $n_{ARC}$ = 1.2<br>$d_{ARC}$ = 100 nm (Top Cell Optimization) | $n_{ARC}$ = 1.2<br>$d_{ARC}$ = 160 nm (Bottom Cell Optimization) |
|---|---|---|
| $\Delta J_{SC\ top}$ | + **4.0%** | + 2.7% |
| $\Delta J_{SC\ bot}$ | + 2.7% | + **3.6%** |

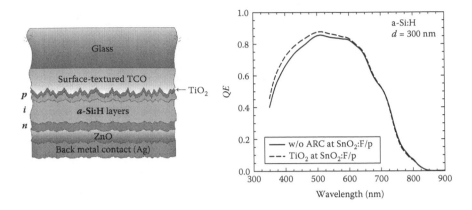

**FIGURE 5.32**   The effect of introduced TiO$_2$ ARC between SnO$_2$:F and p-a-SiC:H layers on *QE* in a single-junction a-Si:H solar cell.

## 5.5   THE DESIGN AND APPLICATION OF ONE-DIMENSIONAL PHOTONIC CRYSTALS FOR THIN-FILM SOLAR CELLS

### 5.5.1   THE BASIC STRUCTURE AND PHYSICS

One-dimensional photonic-crystal (PC) structures are formed by a stack of repeating two (or more) thin layers with a different refractive index and thicknesses in the range of light wavelength (Figure 5.33) (Joannopoulos et al. 2008).

The periodic character of the structure is important in order to obtain the presence and a sufficient amplification of the constructive and deconstructive interferences of electromagnetic waves, originating from multiple reflectances and transmittances of interfaces. The phase relation between the (reflected/transmitted) waves is of importance here. Based on the principle of amplification,

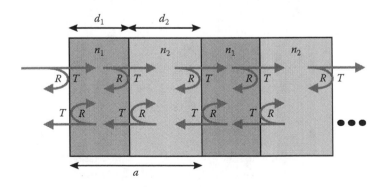

**FIGURE 5.33**   Schematic structure of a 1-D PC.

the electromagnetic waves of particular wavelengths can be entirely reflected from or entirely transmitted through (in the case of non-absorbing layers) the PC structure. The wavelength region, for which the electromagnetic waves cannot propagate throughout the PC, is referred as the *photonic bandgap* (Joannopoulos et al. 2008). The photonic bandgap determines the wavelength region of the ideal reflection (almost 100% can be obtained). The parameters of 1-D PC structures are refractive indexes, $n_1$ and $n_2$ (which can be wavelength dependent) and thicknesses, $d_1$ and $d_2$, of two alternating layers. The period length is defined as $a = d_1 + d_2$. When aiming for a high reflectance (such as in the case of the back reflector), $d_1$ and $d_2$ should each approximately equal a quarter of the wavelength inside the material.

In PC structures we can employ materials that are used in thin-film technology (such as a-Si:H, ZnO, SiO$_x$, SiN$_x$). Also the structures of 1-D PC (i.e., multi-layer stacks) are fully compatible with thin-film fabrication processes. Thus, PC structures are possible candidates for improving light management in thin-film solar cells. In addition to optical requirements, required electrical properties of PC stacks must be met. Conductivity of the stacks can be achieved by using conductive (doped) layers or using appropriate structuring of the PC to enable local contacts of the devices. In this section we focus on optical properties of the PCs. By means of 1-D optical simulations, we demonstrate two possible applications of PCs in thin-film silicon solar cells:

- as wavelength-selective intermediate reflectors in tandem cells and
- as highly reflective back reflectors in thin-film solar cells.

### 5.5.2 ONE-DIMENSIONAL PHOTONIC-CRYSTAL STRUCTURE AS AN INTERMEDIATE REFLECTOR

In Section 5.3.2 we explained the role of an intermediate reflector (interlayer) in tandem solar cells. To improve its basic functionality—to increase reflection at shorter wavelengths and ensure high transmission at longer wavelengths—the idea of realizing a wavelength-selective intermediate reflector is of interest. Introduction of a single-layer interlayer (like ZnO or SiO$_x$) increases the reflectance not only in the short- but also in the long-wavelength part of the spectrum as shown for the case of ZnO interlayer in Figure 5.34 by a dashed line.

In this study we simulated the actual reflectance at the interlayer inside the solar cell structure—reflectance of optical system n-a-Si:H/ZnO/p-μc-Si:H, where n-a-Si:H is the incident medium (*n*-doped layer of the front p-i-n cell) and p-μc-Si:H (*p*-doped layer of the bottom p-i-n cell) is the medium in transmission. Test simulations revealed and experiments confirmed that PC structure consisting of only a few layers can already exhibit pronounced wavelength-selective characteristics. In the structures we used layers that are already present in solar cell structures. Simple testing PC stacks were formed from ZnO and n-a-Si:H layers. The stacks were fabricated on glass substrate at TU Delft.

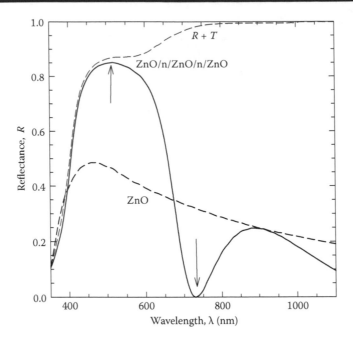

**FIGURE 5.34**    Reflectance of a ZnO interlayer ($d = 100$ nm) and of a wavelength-selective intermediate reflector realized with ZnO ($d_{ZnO} = 70$ nm) and n-a-Si:H layers ($d_{n\text{-}a\text{-}Si:H} = 25$ nm). The reflectance is determined if the layers are inside the solar cell structure (n-a-Si:H incident medium and p-μc-Si:H medium in transmission).

In Figure 5.35, simulations and measurements of one-, two- and three-layer PC stacks are presented.

The thicknesses of the ZnO and n-a-Si:H layers of 130 nm and 25 nm were used, respectively. To be able to compare simulation results with experimental ones, simulations of the optical system outside the solar cell structure (as measured, air/PC/glass/air optical system) were simulated in this case. The PC samples were fabricated at TU Delft. Relatively good agreement between simulated and measured curves is obtained. As for a single ZnO layer on glass, only a weak wavelength dependency is observed; also, outside the solar cell structure, the three-layer ZnO/n-a-Si:H/ZnO stack exhibit pronounced dependency with a high reflectance at shorter ($\lambda < 650$ nm) and low reflectance at longer wavelengths. This gave us motivation to use simulation to design a proper PC stack for the situation inside the cell structure. Thus, we came back to simulation of the n-a-Si:H/intermediate reflector/p-μc-Si:H system, where for the intermediate reflector a PC structure consisting of ZnO and n-a-Si:H layers was applied. We were able to obtain a good wavelength selectivity for the PC in the following configuration: ZnO/n-a-Si:H/ZnO/n-a-Si:H/ZnO with the thickness of ZnO and n-a-Si:H layers of 70 nm and 25 nm, respectively. This presents a PC with two and a half periods in length. The reflectance at shorter wavelengths (e.g., at

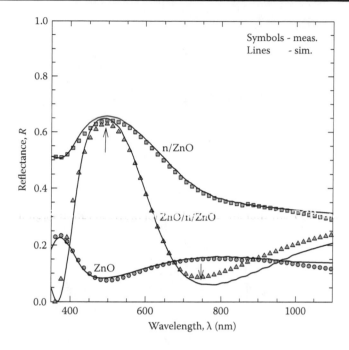

**FIGURE 5.35**    Verification of the concept of a wavelength-selective intermediate reflector: measurement and simulations of the structures deposited on glass and measured in the surrounding air ($d_{ZnO} = 130$ nm, $d_{n\text{-}a\text{-}Si:H} = 25$ nm).

$\lambda \approx 500$ nm) is almost doubled compared to a single ZnO interlayer, whereas at longer wavelengths it is reduced below that of the ZnO interlayer. Such realization of intermediate reflectors still has to be tested in realistic solar cells, especially in those with textured interfaces. Here the purpose is to demonstrate how 1-D optical simulation can be used in the design and testing of different approaches to light manipulation in thin-film solar cell structures.

### 5.5.3    PHOTONIC-CRYSTAL STRUCTURE AS A BACK REFLECTOR

In the case of a back reflector we have to ensure a high reflectance in a sufficiently broad long-wavelength range of the spectrum, depending on the solar cell type. In this way long-wavelength light that is not fully absorbed within its first pass through the absorber layer is efficiently reflected back. Back reflectors in thin-film solar cells are realized mainly by metallic layers such as silver, aluminum, or molybdenum. In the case of thin-film silicon cells, a combination of Al or Ag layers with a thin ZnO TCO layer in front of it is realized, which ensures better reflectivity and hence lower optical losses in the metal layer (Berginski et al. 2008; Dahal et al. 2009). These reflectors (especially Ag-based) exhibit excellent electrical and relatively good optical properties. However, in the case of texturing they suffer from the effect of surface plasmon absorption, which can noticeably reduce their reflectivity characteristics (Springer et al. 2004). Furthermore, there are additional drawbacks related to metal back reflectors, such as relatively high production costs and, if not perfectly sealed, the sensitivity of the contact material to the moisture in the environment (Meier et al. 2005). Therefore, alternative back reflector concepts, such as back reflectors based on PC (this section), or dielectric materials, such as white reflectors (next section), are being investigated.

To demonstrate the possibility of realizations of PC structures with very high broadband reflection by using layers (materials) compatible with the thin-film solar cell technologies, we designed and fabricated a 12-layer (6-period) PC stack with a-SiN$_x$ and i-a-Si:H layers on a glass substrate (Figure 5.36, left).

The thicknesses and the refractive indexes of the layers were optimized by simulations to $d_{\text{a-SiNx}} = 95$ nm, $n_{\text{a-SiNx}} \approx 1.9$ and $d_{\text{i-a-Si:H}} = 50$ nm, $n_{\text{i-a-Si:H}} \approx 3.8$ ($\lambda = 650$ nm). The results of simulated (optical simulator Sun*Shine*) and measured reflectances of the PC stack are presented in Figure 5.36. The PC samples were fabricated at TU Delft. Good agreement is obtained between the predicted (simulated) and measured characteristics. We can observe a high reflectance (close to 100%) in a broad long-wavelength region ($\lambda_0 \approx 650$–950 nm). This wavelength region is important, for example, for the back reflector in thin-film Si solar cells. We can tune the wavelength region of high reflectance by changing the thicknesses or material properties of the layers. In Figure 5.37, besides the measurement of the mentioned 12-layer PC stack (in this figure plotted in gray), the measurements of the same layer structure where the thicknesses of the layers were changed, $d_{\text{a-SiNx}} = 120$ nm and $d_{\text{i-a-Si:H}} = 60$ nm, are presented. For larger thicknesses the wavelength region of high reflectance is shifted to the longer wavelengths ($\lambda_0 = 750$–1100 nm). Simulations indicated that for a broad wavelength range of high reflectance, a layer with a small $n$ and a layer with a

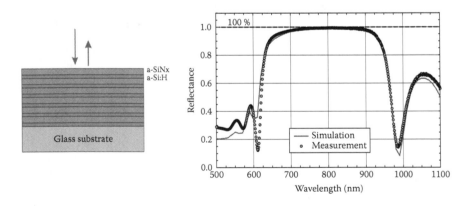

**FIGURE 5.36** A 12-layer photonic crystal realized with a-SiN$_x$ (95 nm) and i-a-Si:H (50 nm) layers deposited on a glass substrate (left). Measured and simulated reflectance of the structure (right).

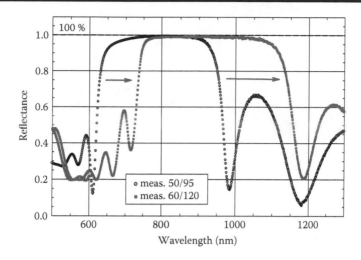

**FIGURE 5.37**  Measured reflectance of the 12-layer PC structures, one with the thicknesses of the a-SiN (95 nm) and i-a-Si:H (50 nm) and the other one with the a-SiN$_x$ (120 nm) and i-a-Si:H (60 nm).

high $n$ are needed. Further on, the absorption in the wavelength region of interest should be low (negligible). Proper thickness tuning and a sufficient number of layers are crucial to the optimal design. Optical modeling is an important tool for the design and optimization of such structures. As mentioned, from the electrical point of view the stacks have to be sufficiently conductive in the vertical direction when including them in solar cell structures.

We used optical simulations to evaluate the potential of applying a simple PC structure as a back reflector in an a-Si:H solar cell. The structure of the cell with a five-layer PC stack at the back is presented in Figure 5.38 (left). The layer thicknesses of the solar cell structure used in the simulations are glass/ZnO:Al (700 nm)/p-a-SiC:H (10 nm)/i-a-Si:H (185 nm)/n-a-Si:H (20 nm)/PC back reflector. The PC attached at the back of the cell is based on the combination of an n-a-Si:H layer (40 nm) and an a-SiO$_x$ (80 nm) layer. Conductivity of both types of layers can be achieved. The first n-a-Si:H layer in the stack is already the $n$-doped layer (20 nm) in the p-i-n solar cell. The results of the simulated absorptances in the i-a-Si:H absorber layer are shown in Figure 5.38. The simulations of the cell with the PC back reflector are represented by a thick curve. The absorptance in the i-a-Si:H layer of the cell with the PC back reflector can be compared to the absorptance of the cell with a metal Ag back reflector (thin full curve). It has to be pointed out that for comparison, the interference fringes of the cells with both back reflectors needed to be matched, which we achieved by using a thicker absorber ($d_{\text{i-a-Si:H}} = 200$ nm) in the case of the cell with the Ag back reflector. By considering this, we can observe that the absorptance in the i-a-Si:H layer in the cell with the simple PC structure already approaches the results obtained with a metal back reflector quite well. The simulation of the cell without any

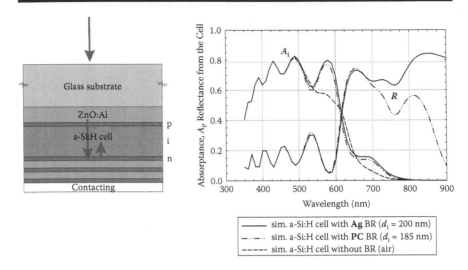

**FIGURE 5.38** Simulated results of absorptance, $A_i$, and reflectance, $R$, for a thin a-Si:H cell with a simple PC back reflector. The simulations for the cell with an Ag back reflector and without back reflector are added for comparison.

back reflector (air) is shown in addition. Further on, the reflectance of the cells with the PC and Ag back reflector are plotted. We can observe that for the long-wavelengths ($\lambda_0 > 700$ nm), the reflectance of the cell with the simple PC drops. This indicates that the simple PC ensures good reflectance just in the required wavelength region for the thin i-a-Si:H absorber; the rest of the long-wavelength light is transmitted through the cell. Experimental verifications of PC as a back reflector in a-Si:H solar cells can be found in Isabella et al. (2009, 2012).

Besides regular (periodic) structures, approaches toward extending the wavelength region of high reflectance have been investigated, as discussed earlier. A possible solution is the modulated PC concept in which the thicknesses and refractive indexes of layers are not fully periodic but modulated with a certain function. As a simple example of a modulated PC structure, a serial combination of two regular (periodic) PC stacks was reported (Krč et al. 2009). The first PC stack exhibits a high reflection for shorter wavelengths while the second PC has highly reflective characteristics for longer wavelengths.

## 5.6  SIMULATION OF DIELECTRIC BACK REFLECTORS—WHITE PAINT

Dielectric back reflectors present an alternative to metal back reflectors, as explained in Section 5.2.5.3. Paints and plastic sheets are of interest in the role of back reflectors because of their simple applicability to thin-film PV technologies (Meier et al. 2005; Kluth et al. 2011). In this section, we show how we can apply 1-D optical modeling to analyze and optimize optical properties of white paint films and complete solar cell devices, including the films as back reflectors. A similar modeling approach can be taken for white sheets or foils.

## 5.6.1    COMPOSITION OF WHITE PAINT AND RESULTS OF OPTICAL CHARACTERIZATION

Basically, the structure of a dry paint film can be represented by a mixture of two primary components: the pigment and the binder. The pigment defines the color of the paint, whereas the binder is used as the matrix in which the pigment particles are dispersed, forming a film. Additionally, other components like volatile substances (carriers) and additives are included in the paint composition to improve its viscosity, drying, and stability characteristics (Goldschmidt and Streitbeger 2003). The external optical properties of dry paint films, however, can be assigned mainly to the pigment and the binder. Therefore, we take only these two components into account in our optical analysis. As the pigment in the commercially available white paints, rutile titanium dioxide ($TiO_2$) particles with the refractive index $n_p = 2.74$ ($\lambda = 550$ nm) and typical particle diameter of 300 nm are most commonly employed (Vargas, Amador, and Niklasson 2006), whereas various polymeric latex-based materials with typical refractive indices $n_b = 1.4$–1.7 are used as the binder (Cotter et al. 1999).

Besides selection of commercially available white paints, we prepared our own compositions for this simulation study. White paint samples were prepared by mixing the rutile $TiO_2$ pigment with a mean diameter of 300 nm and a polymeric binder in a mortar grinder. The pigment volume concentration (*PVC*), which is defined as the ratio of the pigment volume to the entire paint volume, was varied from 5% to 30%. The prepared paints were applied on a borosilicate glass substrate using the "doctor blading" technique (Krebs 2009) and dried out at room temperature (>3 h). The paint film thickness (*d*) of the samples spanned from 14 µm to 150 µm. Additionally, a 100-µm thick sample of the commercially available white paint Tipp-Ex was also prepared as a reference. The white paint samples were used for verification of the model and the optimization trends indicated by the simulations.

The reflectivity and scattering properties of white paint films, which are crucial for efficient back reflectors in thin-film solar cells, we characterize by the total reflectance ($R_{tot}$), the haze parameter for the reflected light ($H_R$), and the angular distribution function of the reflected scattered light ($ADF_R$). In Figure 5.39, the measured $R_{tot}$ and $H_R$ of a commercial white paint Tipp-Ex (100-µm thick film on a glass substrate) are compared with the measurements of a textured Ag reflector (100-nm thick Ag film on $SnO_2$:F Asahi U substrate with $\sigma_{rms} \approx 40$ nm). The results show that $R_{tot}$ of the textured Ag reflector is generally higher than $R_{tot}$ of the white paint sample, which peaks at $\lambda = 440$ nm ($R_{tot} = 0.85$) and then gradually decreases toward longer wavelengths. The abrupt drop of white paint's $R_{tot}$ in the low-wavelength region ($\lambda < 420$ nm) is related to the absorption in $TiO_2$ particles (indirect allowed transitions in $TiO_2$ at about

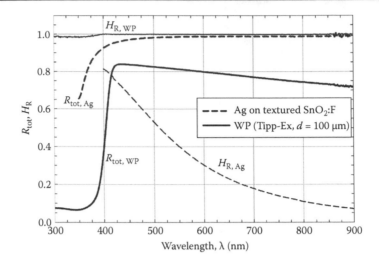

**FIGURE 5.39**    The measured total reflectance and haze parameter of the white paint sample compared to the textured Ag back reflector.

3.02 eV) (Vargas et al. 2000) and is not critical from the back reflector point of view, since the short-wavelength light ($\lambda$ < 550 nm) is efficiently absorbed in the absorber layers of the solar cell before reaching the back reflector. The $H_R$ measurement results in Figure 5.39 show that the white paint sample exhibits extremely high, almost ideal $H_R \approx 1$ throughout the spectral region, whereas $H_R$ of the textured Ag reflector follows exponential decay at longer wavelengths.

The measured *ADF*s of the reflected scattered light for the white paint sample and the textured Ag-based back reflector are shown in Figure 5.40 (measured in a plane perpendicular to the sample, $\lambda = 633$ nm). The results show that compared to the Ag reflector, the white paint sample exhibits much broader *ADF*, which approaches the Lambertian (cosine) distribution (also plotted in Figure 5.40). Similar *ADF* curves were obtained also for other white paint samples of different film thickness and composition (not shown here). These results demonstrate excellent scattering capabilities of white paint (high $H_R$, broad *ADF*), which are far superior to those of conventional metal-based reflectors. However, relatively lower $R_{tot}$ values require further optimization of white paint film composition and thickness. For this purpose, the developed 1-D optical model can be efficiently applied.

### 5.6.2    MODELING OF WHITE PAINT FILMS

Because the paint medium (i.e., the binder with the dispersed pigment particles) is assumed to be isotropic, it is possible to simulate the optical mechanisms taking place within the paint film (propagation of light through the pigment

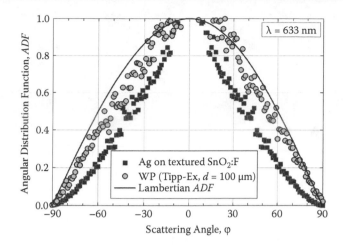

**FIGURE 5.40**   The measured *ADF* (λ = 633 nm) of the white paint sample compared to that of the textured Ag back reflector. The Lambertian (cosine) angular distribution function is also plotted.

and binder materials; reflection, transmission, and scattering at the pigment interfaces) by means of a 1-D optical model. One-dimensional modeling can render good approximation of the measured external optical characteristics of white paint films, such as the total reflectance ($R_{tot}$) and the scattering properties, which are also important for simulations of the entire solar cell structures with white paint back reflectors.

To simulate white paints in the role of back reflectors in thin-film solar cells and to optimize their optical properties, an appropriate optical model is required. Different modeling approaches to white paint's optical properties have been reported, such as the radiative transfer models, for example, the N-flux method (Cotter 1998; Vargas et al. 2000) and the Monte-Carlo method (Nitz et al. 1998). However, the applicability of these models in simulations of complete thin-film solar cells with multilayer structures in which coherent effects of non-scattered light (at the front side of the cell) as well as incoherent propagation of scattered light (at randomly textured interfaces and in white paint films) has not been demonstrated until recently (Lipovšek et al. 2010).

Here we present a 1-D optical modeling approach that enables simulation of the external reflectivity properties of white paint back reflectors and that is fully applicable to 1-D simulations of the entire thin-film solar cells with the existing optical simulators, such as the Sun*Shine* in our case. The structural characteristics of white paint—the thickness of the paint film ($d$), the pigment volume concentration (*PVC*), the refractive index ratio between the pigment and binder materials (*RIR*), and the pigment particle size—are considered as the input parameters of the simulations.

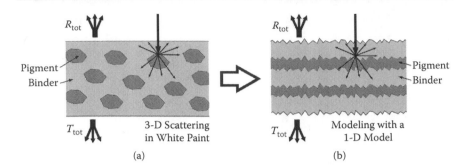

**FIGURE 5.41**   Three-dimensional light scattering in actual white paint (a), and the proposed one-dimensional optical modeling approach (b).

In the developed 1-D optical model, the actual structure of a white paint film is represented by a virtual multilayer structure consisting of two interchanging layers representing the pigment and the binder materials (Figure 5.41). The thickness of the entire multilayer stack is equal to the thickness ($d$) of the paint film, whereas the thickness of each pigment layer ($d_p$) is equal to the average path length of light traveling through a pigment particle (~200 nm for a typical spherical $TiO_2$ particle with a diameter of about 300 nm). The thicknesses of the binder layers and the number of all the layers in the stack are determined to match the *PVC* at certain $d$ of the actual white paint film.

Multiple reflections and transmissions of light at the surfaces of the pigment particles are translated to reflections and transmissions at the interfaces between the layers in the 1-D model. The total reflectance ($R_{tot}$) and the total transmittance ($T_{tot}$) at a pigment/binder interface (including specular and diffused light components) are determined based on the complex refractive indices as at a regular interface in a multilayer structure (see Chapter 2).

The scattering (diffusive) properties of the pigment particles, both in reflection and transmission, are applied to the interfaces between the layers. For this reason, a certain degree of surface nano-texture is introduced at each interface. To define the level of scattering at the interfaces, it was found out that the equations from the scalar scattering theory (Carniglia 1979; Beckmann and Spizzichino 1987), which are commonly employed to determine the level of scattered light at randomly nano-textured interfaces in thin-film solar cells, render a sufficient approximation of the actual scattering of white paint films. Thus, to calculate the haze parameter (ratio of the diffused to total light) for the reflected ($H_R$) and the transmitted light ($H_T$) at an interface in the 1-D model, we used the scalar scattering equations as described in Section 3.2.6.5 [Equations (3.21) and (3.22)]. The $\sigma_{rms}$ parameter in this case can be related mostly to the size of the pigment particles; however, since its direct relation to the average pigment particle diameter has not been derived yet, it is taken as the fitting parameter at this stage. Assuming the previously mentioned relation, the same $\sigma_{rms}$ is considered for white paint compositions having the same average pigment particle diameter. In our simulations, where the average pigment particle diameter was 300 nm, $\sigma_{rms} = 50$ nm was found to be the best fitting solution, rendering good agreement between the simulations and measurements of white paint compositions with different $d$ and $PVC$ values. Simulations revealed that changing $\sigma_{rms}$ does not only affect the level of scattering but also affects the slope of the long-wavelength $R_{tot}$ curve of the white paint film. Furthermore, as the $ADF$ of the scattered light at the textured interfaces (pigment particles) in the 1-D model, the Lambertian (cosine) distribution function was employed both in reflection and transmission, as determined by the angular resolved scattering measurements of paint film samples (see Figure 5.40).

The 1-D model of white paint films, in which the multiple reflection, transmission, scattering, and propagation of light rays are all taken into account, we integrated into the simulator Sun*Shine*. With the presented approach we are able to determine the external optical properties of white paint films of different composition with respect to $d$, $PVC$, and $RIR$. Additionally, by adding the solar cell layers on top of the white paint multilayer stack, the entire solar cell structures with white paint back reflectors can also be simulated, and thus we were able to determine the performance of thin-film amorphous silicon solar cells with different white paint back reflectors.

## 5.6.3   SIMULATION AND OPTIMIZATION OF OPTICAL PROPERTIES OF WHITE PAINT FILMS

In our simulations, the values of the wavelength-dependent refractive index and the absorption coefficient of rutile $TiO_2$ were taken from the literature (average of the values for ordinary and extraordinary rays) (Shenoy and De La Rue 1992; Weber 2002), whereas the refractive index and the absorption coefficient of the polymeric binder used were determined experimentally, employing the reflectance-transmittance method.

To verify the proposed 1-D modeling approach, we simulated the total reflectance of different white paint films and compared the results with the measurements. The input parameters for the simulations were determined according to the properties of the white paint films (pigment and binder materials, paint film thickness, $PVC$), as described next. The $\sigma_{rms}$ parameter, which presents the only fitting parameter, was set to 50 nm in all of the simulations.

First, we verify our simulation approach. The measured (symbols) and the simulated (lines) $R_{tot}$ curves of the selected white paint samples are presented in Figure 5.42. The simulated results are in good agreement with the measurements for all of the samples with different $PVC$ and $d$. A trend of increased $R_{tot}$ as a result of increased $PVC$ and $d$ can be observed, whereas for the films with smaller $R_{tot}$, higher transmission losses can be detected. The $H_R$ parameter, which was calculated from the simulated diffuse reflectance divided by the total reflectance, is close to unity for all the white paint films (not shown here), as observed also in the measurement of the commercially available white paint in Figure 5.39.

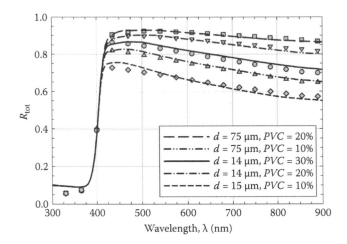

**FIGURE 5.42**   The simulated (lines) and the measured (symbols) total reflectance of a number of white paint samples exhibiting different *d/PVC* combinations.

The level of $R_{tot}$ of a white paint film, as indicated by Figure 5.42, is closely related to the combination of the film thickness ($d$) and the pigment volume concentration ($PVC$). By means of the presented 1-D modeling approach, the role of each of the two parameters can be investigated independently and optimized with respect to higher $R_{tot}$. In these simulations, the pigment and binder materials remain unchanged, and thus $RIR$ is kept constant at 1.83 ($\lambda = 650$ nm). The optimization results are shown for the selected wavelength $\lambda = 650$ nm, representing the long-wavelength region, which is of interest in amorphous and microcrystalline silicon solar cells from the point of view of high back reflection. For the microcrystalline solar cells, longer wavelengths (e.g., $\lambda > 800$ nm) should also be considered. However, by increasing $R_{tot}$ at $\lambda = 650$ nm, $R_{tot}$ at longer wavelengths is also increased, as indicated by the simulations.

In Figure 5.43, the simulated $R_{tot}$ of white paint films with four different $PVCs$ is presented as a function of film thickness. The simulations are again in good agreement with measured points, confirming the optimization trends obtained by the 1-D modeling. All four plots in Figure 5.43 indicate similar behavior of $R_{tot}$, which increases dramatically as $d$ increases from 10 µm to 100 µm. This can be explained by the fact that the thicker the film is, the more pigment/binder interfaces there are in the coating (in the actual film as well as in the 1-D representation). Subsequently, with the increased number of interfaces, the cumulative $R_{tot}$ is also increased since the light can be reflected back more times. However, because of the light absorption in the paint film (both the pigment and especially the binder are absorptive in the long-wavelength region), a saturation of $R_{tot}$ below the ideal value $R_{tot} = 1$ occurs at a certain film thickness. Modeling can be used to determine this saturation region for the paints with

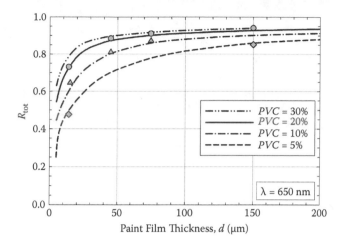

**FIGURE 5.43**   The simulated results (lines, see legend) and the measured results (symbols) of total reflectances as a function of paint film thickness. Different lines and symbols correspond to different pigment volume concentrations of the films (measurements: circles — $PVC = 20\%$; triangles — $PVC = 10\%$; and rotated squares — $PVC = 5\%$)

different $PVC$s, and thus we can evaluate the minimal required thickness of the film that still renders the highest $R_{tot}$. Simulations indicated that if the absorption losses in the white paint film were smaller, the optimal thickness would shift toward larger values, and higher maximal $R_{tot}$ would also be achieved (effect not shown here).

The simulated $R_{tot}$ of white paint films as a function of $PVC$ is presented in Figure 5.44 for four different film thicknesses. The trends observed in the graph are in accordance with those indicated by Figure 5.43. As the $PVC$ increases from 5% to 30%, $R_{tot}$ of the paint film is increased substantially. This can be attributed to the fact that as the $PVC$ is increased, the number of the pigment/binder interfaces in the same paint film volume is increased as well (also in the 1-D representation), leading to higher cumulative $R_{tot}$. The simulations reveal that saturation of $R_{tot}$ appears at $PVC > 40\%$, indicating the optimal pigment volume concentration for a certain paint film thickness. It should be noted, however, that the phenomenon of "crowding" (agglomeration) of the pigment, which can start to take place in the paint films at higher $PVC$ values (>50%), was not included in our simulations. Such agglomeration leads to a reduction of the (effective) pigment/binder interfaces and consequently to a decrease in $R_{tot}$. Furthermore, structural stability issues may also arise in films with high $PVC$ (Goldschmidt and Streitbeger 2003).

In all of the preceding simulations, the refractive index ratio of the paint was kept constant at $RIR = 1.83$ ($n_p = 2.69$, $n_b = 1.47$ at $\lambda = 650$ nm). However, because this ratio dictates the reflectance at a single pigment/binder interface according

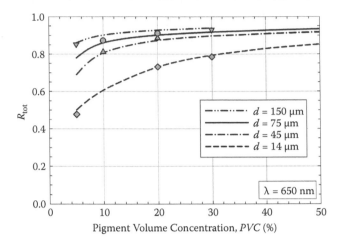

**FIGURE 5.44**   The simulated results (lines, see legend) and the measured results (symbols) of total reflectances as a function of pigment volume concentration. Different lines and symbols correspond to different thicknesses of the films (measurements: triangles down — $d$ = 150 μm; circles — $d$ = 75 μm; triangles up— $d$ = 45 μm; and rotated squares — $d$ = 14 μm).

to Equation (5.4), it also presents one of the influencing parameters that affect $R_{tot}$ of the paint film (besides $d$ and $PVC$).

$$R_{pigment/binder} \approx \left( \frac{n_p - n_b}{n_p + n_b} \right)^2 = \left( \frac{n_p/n_b - 1}{n_p/n_b + 1} \right)^2 = \left( \frac{RIR - 1}{RIR + 1} \right)^2 \qquad (5.4)$$

   In our simulation study, $RIR$ was varied from 1.0 to 2.6 by changing the refractive index of the pigment material ($n_b$ was kept constant at 1.47). In Figure 5.45, the results at the selected wavelength $\lambda$ = 650 nm for the case of a white paint film with $d$ = 50 μm and $PVC$ = 20% are presented. Besides the simulated $R_{tot}$ of the paint film, $R_{tot}$ corresponding to the reflectance of a single binder/pigment interface [calculated from Equation (5.4)] is also shown. Simulations indicate that $RIR$ plays a crucial role in affecting $R_{tot}$. The $R_{tot}$ of the white paint increases noticeably with the refractive index ratio and reaches saturation at high $RIR$ values (>2.2). Here, the increase in $R_{tot}$ is no longer related to the number of pigment/binder interfaces, but instead to the reflectance at a single interface, which increases with increasing $RIR$. In Figure 5.44, the refractive index ratios in the case of three common white pigment materials—rutile $TiO_2$, anatase $TiO_2$, and ZnO (the corresponding refractive indices at 650 nm are 2.69, 2.55, and 2.02, respectively)—are indicated by vertical lines. It is evident that among these three, the rutile $TiO_2$-based pigment presents the best choice for achieving high $R_{tot}$, whereas ZnO is the least favorable because of its lowest refractive index. Alternatively,

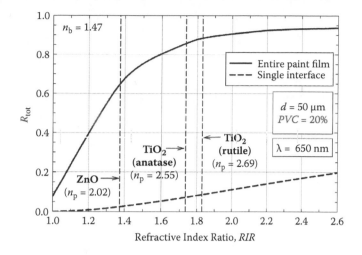

**FIGURE 5.45**   The simulated total reflectance of a 50-µm thick white paint film with *PVC* = 20% as a function of *RIR*. Also plotted is the reflectance at a single rutile TiO$_2$ pigment/binder interface as a function of *RIR* (dashed line). The *RIR*s in the case of three typical white pigments are indicated with vertical lines (the same binder with $n_b$ = 1.47 is assumed in all three cases).

instead of changing the pigment material, binders with lower $n_b$ could also be employed to increase *RIR* and thus boost the total reflectance of the white paint film.

### 5.6.4   SIMULATION OF SOLAR CELLS WITH WHITE PAINT FILMS

We simulated thin-film amorphous silicon (*a*-Si:H) p-i-n solar cells on two types of TCO superstrates: (i) glass/ZnO:Al with flat surface, and (ii) glass/SnO$_2$:F (Asahi U) with a natural surface texture ($\sigma_{rms} \approx 40$ nm). Both front TCO films were of the same thickness (800 nm). The thickness of the intrinsic i-a-Si:H absorber layer was 300 nm. To form the back electrical contact of the cell with white paint back reflectors, a 350-nm thick layer of ZnO:Al was considered between the n-a-Si:H layer and white paint films back reflectors. The cells in this structure were also fabricated at TU Delft. To demonstrate the improvements related to the optimization of white paint back reflectors, external quantum efficiencies (*EQE*) of the a-Si:H solar cells were simulated and measured. The short-circuit current densities ($J_{SC}$) of the cells were calculated from the *EQE* curves by considering the AM 1.5 spectrum.

The simulated (a) and the measured (b) long-wavelength ($\lambda > 550$ nm) *EQE* curves of a-Si:H solar cells deposited on flat ZnO:Al and textured SnO$_2$:F TCO superstrates are shown in Figures 5.46 and 5.47, respectively. At shorter

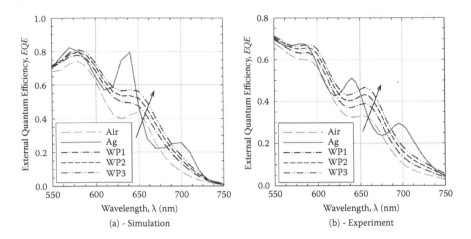

**FIGURE 5.46**   Simulated (a) and measured (b) external quantum efficiencies of thin-film a-Si:H solar cells on a flat ZnO:Al superstrate with different back reflectors. Note that the *EQE* axis in the plot showing the measured results is rescaled for easier comparison of the trends.

wavelengths the optical properties of back reflectors do not affect the *EQE*, since the short-wavelength light is efficiently absorbed (due to high absorption coefficients) before reaching the back side of the cell. Three types of white paint back reflectors were employed in the analysis, labeled WP1 ($d = 14$ µm, $PVC = 5\%$), WP2 ($d = 14$ µm, $PVC = 20\%$), and WP3 ($d = 150$ µm, $PVC = 30\%$).

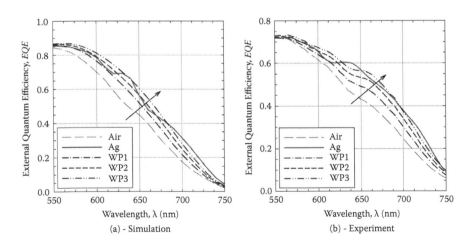

**FIGURE 5.47**   Simulated (a) and measured (b) external quantum efficiencies of thin-film a-Si:H solar cells on a textured SnO$_2$:F superstrate with different back reflectors. Note that the *EQE* axis in the plot showing the measured results is rescaled for easier comparison of the trends.

The *EQE* of the cell without any back reflector after the ZnO:Al layer (air surrounding) and the *EQE* of the cell with an Ag reflector deposited on top of the ZnO:Al layer are shown for comparison. These results demonstrate the effects of white paint optimization on the performance of the solar cells. As $R_{tot}$ of the white paint back reflector is increased (by increasing $d$ and *PVC* of the film), the long-wavelength *EQE* of the cell is boosted notably in both types of solar cells deposited on flat and textured superstrates. Since in the case of cells deposited on flat superstrate the light is scattered only at the white paint reflector, some interference fringes as a consequence of the (coherent) specular light can still be observed in *EQE*. The measured *EQE*s confirm the trends observed in the simulated *EQE*s. However, the measured *EQE* values are generally lower than the simulated ones (note different *EQE* scales), which we attribute mainly to the non-optimized optical and electrical characteristics of the back low-temperature ZnO:Al TCO film in the fabricated cells. Nevertheless, the same trends are expected also in electrically optimized solar cells.

The comparison of $J_{SC}$ values that were calculated from the simulated long-wavelength *EQE* reveals an increase of 10.5% and 10.7% if the WP3 reflector is employed instead of WP1 in the cell deposited on the flat and the textured superstrate, respectively. The absolute $J_{SC}$ values calculated from *EQE* at $\lambda > 550$ nm for the cells with WP3 back reflector are 6.35 mA/cm$^2$ (flat) and 7.38 mA/cm$^2$ (textured superstrate). Considering the wavelength spectrum from 350 to 750 nm, the corresponding $J_{SC}$s of the cells with WP3 back reflector are 12.25 mA/cm$^2$ (flat) and 14.66 mA/cm$^2$ (textured superstrate). These values reach and even exceed the $J_{SC}$ values obtained with the Ag back reflector (12.11 mA/cm$^2$ for the flat and 14.51 mA/cm$^2$ for the textured superstrate). The same relative trends in $J_{SC}$ as determined from the simulations can also be obtained from the measured *EQE*s.

As shown in this chapter, 1-D optical modeling and simulations can be applied to various optical approaches to thin-film photovoltaic devices and can even be used to describe more complicated effects, for instance, the plasmonic behavior of metal nano-particles in thin-film solar cells (Schmid et al. 2011). Sometimes we use 1-D modeling along with other, more complex 3-D rigorous simulations. They are fast, many effects are taken into account, and they are important for our understanding. We recommend that beginners to modeling start with at least a few simulations in 1-D space to get a proper feeling for optics in thin-film devices, such as effect of thickness variation, introducing scattering, and so on.

## REFERENCES

Beckmann, P., and A. Spizzichino. 1987. *The Scattering of Electromagnetic Waves from Rough Surfaces.* Artech Print on Demand.

Berginski, M., J. Hüpkes, A. Gordijn, W. Reetz, T. Wätjen, B. Rech, and M. Wuttig. 2008. "Experimental Studies and Limitations of the Light Trapping and Optical Losses in Microcrystalline Silicon Solar Cells." *Solar Energy Materials and Solar Cells* 92 (9): 1037–1042. doi:10.1016/j.solmat.2008.03.005.

226     *References*

Buehlmann, P., J. Bailat, D. Dominé, A. Billet, F. Meillaud, A. Feltrin, and C. Ballif. 2007. "In Situ Silicon Oxide Based Intermediate Reflector for Thin-Film Silicon Micromorph Solar Cells." *Applied Physics Letters* 91 (14) (October 1). doi:10.1063/1.2794423.
Carniglia, C. K. 1979. "Scalar Scattering Theory for Multilayer Optical Coatings." *Optical Engineering* 18: 104–115.
Cotter, J. E. 1998. "Optical Intensity of Light in Layers of Silicon with Rear Diffuse Reflectors." *Journal of Applied Physics* 84 (1): 618–624. doi:10.1063/1.368065.
Cotter, J. E, R. B. Hall, M. G. Mauk, and A. M. Barnett. 1999. "Light Trapping in Silicon-Film™ Solar Cells with Rear Pigmented Dielectric Reflectors." *Progress in Photovoltaics: Research and Applications* 7 (4): 261–274. doi:10.1002/(SICI)1099-159X(199907/08)7:4<261::AID-PIP256>3.0.CO;2-A.
Dahal, L. R., D. Sainju, J. Li, N. J. Podraza, M. N. Sestak, and R. W. Collins. 2009. "Comparison of Al/ZnO and Ag/ZnO Interfaces of Back-Reflectors for Thin Film Si:H Photovoltaics." 34th IEEE Photovoltaic Specialists Conference (PVSC), June 7–12, 2009, Philadelphia, Pennsylvania, IEEE, pp. 1702–1707. doi:10.1109/PVSC.2009.5411428.
Das, C., A. Lambertz, J. Hüpkes, W. Reetz, and F. Finger. 2008. "A Constructive Combination of Antireflection and Intermediate-Reflector Layers for a-Si/mu c-Si Thin Film Solar Cells." *Applied Physics Letters* 92 (5) (February 4). doi:10.1063/1.2841824.
Dominé, D., P. Buehlmann, J. Bailat, A. Billet, A. Feltrin, and C. Ballif. 2008. "Optical Management in High-Efficiency Thin-Film Silicon Micromorph Solar Cells with a Silicon Oxide Based Intermediate Reflector." *Physica Status Solidi–Rapid Research Letters* 2 (4) (August): 163–165. doi:10.1002/pssr.200802118.
Dullweber, T., G. Hanna, W. Shams-Kolahi, A. Schwartzlander, M. A. Contreras, R. Noufi, and H. W. Schock. 2000. "Study of the Effect of Gallium Grading in Cu(In,Ga)Se₂." *Thin Solid Films* 361–362 (February): 478–481. doi:10.1016/S0040-6090(99)00845-7.
Fay, S., J. Steinhauser, N. Oliveira, E. Vallat-Sauvain, and C. Ballif. 2007. "Opto-electronic Properties of Rough LP-CVD ZnO:B for Use as TCO in Thin-Film Silicon Solar Cells." *Thin Solid Films* 515 (24) (October 15): 8558–8561. doi:10.1016/j.tsf.2007.03.130.
Fischer, D., S. Dubail, J. A. A. Selvan, N. P. Vaucher, R. Platz, C. Hof, U. Kroll, et al. 1996. "The 'Micromorph' Solar Cell: Extending a-Si:H Technology Towards Thin Film Crystalline Silicon." Conference Record of the 25th IEEE Photovoltaic Specialists Conference, May 13–17, 1996, Washington, DC, IEEE, pp. 1053–1056. doi:10.1109/PVSC.1996.564311.
Fujibayashi, T., T. Matsui, and M. Kondo. 2006. "Improvement in Quantum Efficiency of Thin Film Si Solar Cells due to the Suppression of Optical Reflectance at Transparent Conducting Oxide/Si Interface by TiO2/ZnO Antireflection Coating." *Applied Physics Letters* 88 (18) (May 1). doi:10.1063/1.2200741.
Goldschmidt, A., and H.-J. Streitbeger. 2003. *BASF Handbook on Basics of Coating Technology*. Vincentz.
Green, M. A. 2005. *Third Generation Photovoltaics: Advanced Solar Energy Conversion*. Springer.
Guha, S., and J. Yang. 2010. "Thin Film Silicon Photovoltaic Technology: From Innovation to Commercialization." *MRS Online Proceedings Library* 1245. doi:10.1557/PROC-1245-A01-01.
Hongsingthong, A., T. Krajangsang, I. A. Yunaz, S. Miyajima, and M. Konagai. 2010. "ZnO Films with Very High Haze Value for Use as Front Transparent Conductive Oxide Films in Thin-Film Silicon Solar Cells." *Applied Physics Express* 3 (5). doi:10.1143/APEX.3.051102.
Isabella, O., A. Čampa, M. Heijna, W. J. Soppe, A. J. M. van Erven, R. H. Franken, H. Borg, and M. Zeman. 2008. "Diffraction Gratings for Light Trapping in Thin-Film Silicon Solar Cells." 25th European Photovoltaic Solar Energy Conference

and Exhibition / 5th World Conference on Photovoltaic Energy Conversion, September 6–10, 2010, Valencia, Spain. WIP-Renewable Energies, pp. 2320–2324.

Isabella, O., S. Dobrovolskiy, G. Kroon, M. Zeman. 2012. "Design and Application of Dielectric Distributed Bragg Back Reflector in Thin-Film Silicon Solar Cells," *Journal of Non-Crystalline Solids*, 358 (17 September): 2295–2298. ISSN 0022-3093, 10.1016/j.jnoncrysol.2011.11.025.

Isabella, O., J. Krč, and M. Zeman. 2010. "Modulated Surface Textures for Enhanced Light Trapping in Thin-Film Silicon Solar Cells." *Applied Physics Letters* 97 (10) (September 6). doi:10.1063/1.3488023.

Isabella, O., B. Lipovšek, J. Krč, and M. Zeman. 2009. "Photonic Crystal Back Reflector in Thin-Film Silicon Solar Cells." *MRS Online Proceedings Library* 1153: 1153–A03–05. doi:10.1557/PROC-1153-A03-05.

Joannopoulos, J. D., S. G. Johnson, J. N. Winn, and R. D. Meade. 2008. *Photonic Crystals: Molding the Flow of Light.* 2nd ed. Princeton University Press.

Kambe, M., M. Fukawa, N. Taneda, Y. Yoshikawa, K. Sato, K. Ohki, S. Hiza, A. Yamada, and M. Konagai. 2003. "Improvement of Light-Trapping Effect on Microcrystalline Silicon Solar Cells by Using High Haze Transparent Conductive Oxide Films." 3rd World Conference on Photovoltaic Energy Conversion, May 11–18, 2003, Osaka, Japan, IEEE, pp. 1812–1815.

Kluth, O, B. Rech, L. Houben, S. Wieder, G. Schöpe, C. Beneking, H. Wagner, A. Löffl, and H. W. Schock. 1999. "Texture Etched ZnO:Al Coated Glass Substrates for Silicon Based Thin Film Solar Cells." *Thin Solid Films* 351 (1–2): 247–253. doi:10.1016/S0040-6090(99)00085-1.

Kluth, O., J. Kalas, M. Fecioru-Morariu, P. A. Losio, and J. Hötzel. 2011. "The Way to 11% Stabilized Module Efficiency Based on 1.4m2 Micromorph˚ Tandem." In 2354–2357. Hamburg, Germany.

Krč, J., F. Smole, and M. Topič. 2006. "Advanced Optical Design of Tandem Micromorph Silicon Solar Cells." *Journal of Non-Crystalline Solids* 352 (9–20): 1892–1895. doi:10.1016/j.jnoncrysol.2005.12.040.

Krč, J., M. Zeman, O. Kluth, E. Smole, and M. Topič. 2003. "Effect of Surface Roughness of ZnO : Al Films on Light Scattering in Hydrogenated Amorphous Silicon Solar Cells RID A-5194-2008." *Thin Solid Films* 426 (1-2) (February 24): 296–304. doi:10.1016/S0040-6090(03)00006-3.

Krč, J., M. Zeman, S. L. Luxembourg, and M. Topič. 2009. "Modulated Photonic-Crystal Structures as Broadband Back Reflectors in Thin-Film Solar Cells." *Applied Physics Letters* 94 (15): 153501. doi:10.1063/1.3109781.

Krč, J., M. Zeman, F. Smole, and M. Topič. 2002. "Optical Modeling of a-Si:H Solar Cells Deposited on Textured Glass/SnO$_2$ Substrates RID A-5194-2008." *Journal of Applied Physics* 92 (2) (July 15): 749–755. doi:10.1063/1.1487910.

Krebs, F. C. 2009. "Fabrication and Processing of Polymer Solar Cells: A Review of Printing and Coating Techniques." *Solar Energy Materials and Solar Cells* 93 (4): 394–412. doi:10.1016/j.solmat.2008.10.004.

Lipovšek, B. 2012. Advanced light management in thin-film optoelectronic devices, PhD Thesis, University of Ljubljana, Slovenia.

Lipovšek, B., J. Krč, O. Isabella, M. Zeman, and M. Topič. 2010. "Modeling and Optimization of White Paint Back Reflectors for Thin-Film Silicon Solar Cells." *Journal of Applied Physics* 108 (10): 103115. doi:10.1063/1.3512907.

Lipovšek, B., J. Krč, and M. Topič. 2010. "Potential of Advanced Optical Concepts in Chalcopyrite-based Solar Cells." E-MRS, Symposium B, Strasbourg, June 8–12, 2009. Proceedings of Inorganic and Nanostructured Photovoltaics : E-MRS 2009, Symposium B, (*Energy Procedia*, vol. 2, no. 1, pp. 143–150, published 2010), Elsevier.

Meier, J., U. Kroll, J. Spitznagel, S. Benagli, T. Roschek, G. Pfanner, C. Ellert, et al. 2005. "Progress in Up-scaling of Thin Film Silicon Solar Cells by Large-Area PECVD KAI Systems." *Conference Record of the 31st IEEE Photovoltaic Specialists Conference*, Jan 3–7, 2005, Orlando, Florida, IEEE, pp. 1464–1467. doi:10.1109/PVSC.2005.1488418.

Meier, J., J. Spitznagel, U. Kroll, C. Bucher, S. Faÿ, T. Moriarty, and A. Shah. 2004. "Potential of Amorphous and Microcrystalline Silicon Solar Cells." *Thin Solid Films* 451–452: 518–524. doi:10.1016/j.tsf.2003.11.014.

Nishiwaki, S., S. Siebentritt, P. Walk, and M. C. Lux-Steiner. 2003. "A Stacked Chalcopyrite Thin-Film Tandem Solar Cell with 1.2 V Open-Circuit Voltage." *Progress in Photovoltaics: Research and Applications* 11 (4) (June 1): 243–248. doi:10.1002/pip.486.

Nitz, P., J. Ferber, R. Stangl, H. Rose Wilson, and V. Wittwer. 1998. "Simulation of Multiply Scattering Media." *Solar Energy Materials and Solar Cells* 54 (1–4): 297–307.

Oyama, T., M. Kambe, N. Taneda, and K. Masumo. 2008. "Requirements for TCO Substrate in Si-based Thin Film Solar Cells: Toward Tandem." *MRS Online Proceedings Library* 1101: 1101–KK02–01. doi:10.1557/PROC-1101-KK02-01.

Pieters, B. E., J. Krč, and M. Zeman. 2006. "Advanced Numerical Simulation Tool for Solar Cells: ASA5." *IEEE 4th World Conference on Photovoltaic Energy Conversion*, Waikoloa, Hawaii, May 7–12, 2006, IEEE. pp.1513–1516. doi:10.1109/WCPEC.2006.279758.

Repins, I., M. A. Contreras, B. Egaas, C. DeHart, J. Scharf, C. L. Perkins, B. To, and R. Noufi. 2008. "19.9%-Efficient ZnO/CdS/CuInGaSe₂ Solar Cell with 81.2% Fill Factor." *Progress in Photovoltaics: Research and Applications* 16 (3) (May 1): 235–239. doi:10.1002/pip.822.

Schmid, M., R. Klenk, M. C. Lux-Steiner, M. Topič, and J. Krč. 2011. "Modeling Plasmonic Scattering Combined with Thin-Film Optics." *Nanotechnology* 22 (2) (January 14). doi:10.1088/0957-4484/22/2/025204.

Schmid, M., J. Krč, R. Klenk, M. Topič, and M. C. Lux-Steiner. 2009. "Optical Modeling of Chalcopyrite-based Tandems Considering Realistic Layer Properties." *Applied Physics Letters* 94 (5): 053507. doi:10.1063/1.3077613.

Selvan, J. A. A., A. E. Delahoy, S. Guo, and Y.-M. Li. 2006. "A New Light Trapping TCO for nc-Si:H Solar Cells." *Solar Energy Materials and Solar Cells* 90 (18–19) (November 23): 3371–3376. doi:10.1016/j.solmat.2005.09.018.

Shenoy, M. R, and R. M. De La Rue. 1992. "On the Refractive Index of Rutile." *IEE Proceedings Journal* 139 (2): 163–165.

Söderström, T., F.-J. Haug, X. Niquille, V. Terrazzoni, and C. Ballif. 2009. "Asymmetric Intermediate Reflector for Tandem Micromorph Thin Film Silicon Solar Cells." *Applied Physics Letters* 94 (6): 063501. doi:10.1063/1.3079414.

Springer, J., A. Poruba, L. Müllerova, M. Vaneček, O. Kluth, and B. Rech. 2004. "Absorption Loss at Nanorough Silver Back Reflector of Thin-Film Silicon Solar Cells." *Journal of Applied Physics* 95 (3): 1427–1429. doi:10.1063/1.1633652.

Springer, J., A. Poruba, and M. Vaneček. 2004. "Improved Three-Dimensional Optical Model for Thin-Film Silicon Solar Cells." *Journal of Applied Physics* 96 (9): 5329. doi:10.1063/1.1784555.

Staebler, D., and C. Wronski. 1977. "Reversible Conductivity Changes in Discharge-Produced Amorphous Si." *Applied Physics Letters* 31 (4): 292–294. doi:10.1063/1.89674.

Vargas, W. E., A. Amador, and G. A. Niklasson. 2006. "Diffuse Reflectance of TiO₂ Pigmented Paints: Spectral Dependence of the Average Pathlength Parameter and the Forward Scattering Ratio." *Optics Communications* 261 (1): 71–78. doi:10.1016/j.optcom.2005.11.059.

Vargas, W. E., P. Greenwood, J. E. Otterstedt, and G. A. Niklasson. 2000. "Light Scattering in Pigmented Coatings: Experiments and Theory." *Solar Energy* 68 (6): 553–561.

Weber, M. J. 2002. *Handbook of Optical Materials.* CRC Press.

Xu, X., J. Zhang, D. Beglau, T. Su, S. Ehlert, Y. Li, G. Pietka, et al. 2010. "High Efficiency Large-Area a-SiGe:H and nc-Si:H Based Multi-Junction Solar Cells: A Comparative Study." 25th European Photovoltaic Solar Energy Conference and Exhibition / 5th World Conference on Photovoltaic Energy Conversion, September 6–10, 2010, Valencia, Spain. WIP-Renewable Energies, pp. 2783–2787.

Yamamoto, K., M. Yoshimi, Y. Tawada, S. Fukuda, T. Sawada, T. Meguro, H. Takata, et al. 2002. "Large Area Thin Film Si Module." *Solar Energy Materials and Solar Cells* 74 (1–4) (October): 449–455. doi:10.1016/S0927-0248(02)00113-7.

Young, D. L., J. Keane, A. Duda, J. A. M. AbuShama, C. L. Perkins, M. Romero, and R. Noufi. 2003. "Improved Performance in ZnO/CdS/CuGaSe$_2$ Thin-Film Solar Cells." *Progress in Photovoltaics: Research and Applications* 11 (8) (December 1): 535–541. doi:10.1002/pip.516.

Zeman, M., J. A. Willemen, L. L. A. Vosteen, G. Tao, and J. W. Metselaar. 1997. "Computer Modelling of Current Matching in a-Si:H/a-Si:H Tandem Solar Cells on Textured TCO Substrates." *Solar Energy Materials and Solar Cells* 46 (2) (May): 81–99. doi:10.1016/S0927-0248(96)00094-3.

Zeman, M., R. A. C. M. M. van Swaaij, J. W. Metselaar, and R. E. I. Schropp. 2000. "Optical Modeling of *a*-Si:H Solar Cells with Rough Interfaces: Effect of Back Contact and Interface Roughness." *Journal of Applied Physics* 88 (11): 6436. doi:10.1063/1.1324690.

# Two-Dimensional Optical Simulations

6

## 6.1   INTRODUCTION

In Chapter 4 we presented and discussed aspects of two-dimensional (2-D) and three-dimensional (3-D) optical modeling and showed the development of 2-D optical simulator FEMOS 2-D based on the finite element method. In this chapter we demonstrate examples of 2-D simulations of thin-film silicon solar cells, using the FEMOS 2-D simulator. One can use any other 2-D numerical simulator, with similar characteristics, to carry out the presented simulations.

As mentioned earlier, 2-D simulations enable the inclusion of geometrical variations of any surface nano-texture (random or periodic) in one lateral direction. Such direct relation between surface morphology, solar cell structure, and output characteristics of PV devices on the other side appears to be very convenient for thin-film solar cells in which different interface textures are of interest to researchers aiming to increase output performance. Suitable for simulations, but also showing high potential for controlled light scattering, are periodic one-dimensional (1-D) surface textures (diffraction gratings, see also Chapter 4). As explained, due to periodic character, only one or even one half of the period of lateral variation is sufficient to be included in the simulation, reducing the number of simulation nodes and computational time substantially. Therefore, we can still carry out simulations in a reasonable time (hours or even minutes). Periodic textures can be realized on rigid (glass) or, even more attractive, on flexible substrates, by using special coatings (dedicated lacquers) on which periodic textures are embossed by thermoplastic or ultra-violet techniques (Battaglia et al. 2011; K. Söderström et al. 2011; Soppe et al. 2011).

Different groups have been using 2-D simulations to investigate the optical situation in thin-film silicon solar cells with periodic textures introduced at the interfaces of the structures (Haase and Stiebig 2006, 2007; Čampa 2009; Lipovšek et al. 2010; Dewan et al. 2011). Here, we present some demonstrative examples of simulations with the FEMOS 2-D simulator. We investigate anti-reflective and scattering properties of 1-D periodic textures and perform the optimization of the texture parameters for different types of thin-film silicon solar cells. In addition, we present simulations of the photonic-crystal structure with periodically textured interfaces.

## 6.2   APPLICATION OF ONE-DIMENSIONAL PERIODIC TEXTURES IN THIN-FILM SOLAR CELLS

### 6.2.1   BASIC PROPERTIES OF DIFFRACTION GRATINGS

Diffraction gratings are optical elements with a precise pattern of microscopic periodic structures. The pattern in 1-D gratings is a corrugated surface of aligned grooves (no changes in surface in the direction of grooves), while 2-D gratings are formed by the periodic variation of the surface in both Cartesian directions of the surface plain. At an incidence of light at diffraction grating,

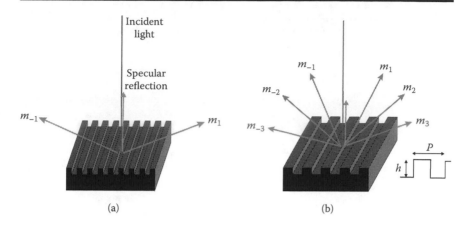

**FIGURE 6.1**    Schematic of light scattering in discrete angles (diffraction orders *m*) at a diffraction grating with smaller (a) and larger (b) period.

scattering of light into discrete modes (orders, angles) occurs as illustrated in Figure 6.1. The values of discrete angles depend on the period, *P*, of the lateral perturbations of the surface textures and can be defined by the grating equation (Beckmann and Spizzichino 1987) [Equation (6.1)]:

$$\varphi_{scatt\,m} = \arcsin\left(\frac{m \cdot \lambda}{n \cdot P} - \sin\varphi_{inc}\right) \tag{6.1}$$

where *m* represents the number of the diffraction order (positive or negative), $\varphi_{scat\,m}$ is the discrete angle of the *m*-th order, $\lambda$ is the wavelength of light, *n* is the refractive index of the incident medium, *P* is the grating period, and $\varphi_{inc}$ is the incident angle of applied illumination.

While the period determines the angles, the height and the shape of the grating affect the intensity of diffracted rays. Analytical approximations determine the scattering intensities of diffracted rays for different types of gratings (Beckmann and Spizzichino 1987); however, numerical modeling can be used to define the entire optical situation.

To demonstrate the scattering abilities of a 1-D grating we show the measured haze parameter *(H)* and the angular distribution function *(ADF)* of selected gratings in Figure 6.2. The measurement was done in surrounding air. Figure 6.2 shows that high haze (close to unity) and the *ADF* with intensified scattering in discrete large angles can be achieved. Such optical characteristics are desired for efficient light trapping in thin-film solar cells. Optical properties can be optimized according to requirements of thin-film solar cells by changing the *P*, *h*, and the shape of the grating. Diffraction gratings (1-D or 2-D) are of interest for testing and implementation in thin-film solar cells as a periodically textured superstrate (front side) or substrate (back side). In simulations, the

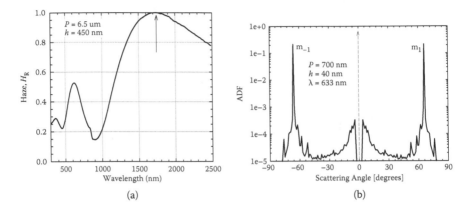

(a)                                                    (b)

**FIGURE 6.2**    Measured haze parameter of reflected light at an aluminum grating (a) and measured *ADF* of transmitted light of the glass/periodically textured lacquer/air system (b). High haze can be achieved in a specific wavelength region as indicated by the arrow in (a), as well as scattering in broad angles as shown in (b). The peaks in the *ADF* are present where positive and negative first scattering order appears. The specular reflected beam (denoted with a dashed arrow) is not present in the measurements.

situation inside the solar cell structure has to be considered rather than the far-field scattering properties of the gratings as measured in air. In this respect 2-D optical simulation based on rigorous solving of a wave equation is a perfect tool for investigation and optimization of the texture, where vertical structure (layers) and lateral variations (texture) are incorporated directly in the description of the structure geometry. Such simulation of the electro-magnetic situation includes also the optical effects related to the near field, which is important when the thicknesses of layers are in the range of light wavelengths as is the case in thin-film solar cells. In the next section, different optical effects related to 1-D diffraction gratings in thin-film solar cells are analyzed by means of the FEMOS 2-D simulator.

### 6.2.2   ANTIREFLECTION AND LIGHT SCATTERING AT PERIODICALLY TEXTURED INTERFACES IN THIN-FILM SILICON SOLAR CELLS

As mentioned in Section 5.4, Chapter 5, antireflecting coatings or (nano-)textures can be used to reduce the reflection of light at front interfaces of photovoltaic devices. We present results of a few simulation cases of different periodic textures and their influence on the total reflectance, $R_{tot}$, of the textured interface. In particular, we investigate the ZnO:Al TCO/p-μc-Si:H interface as one of the front interfaces in μc-Si:H solar cells, having relative high reflection. We applied triangular and rectangular gratings (Figure 6.3, left) as a periodic texture

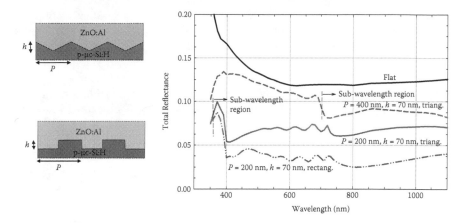

**FIGURE 6.3**  Simulated 1-D grating structures (right) and results of the total reflectance of the ZnO:Al/p-µc-Si:H interface (left). The gratings with triangular and rectangular profile and two different periods are applied to the interface. The anti-reflecting effect is present and enhanced, especially in the sub-wavelength region of periods ($P < \lambda_0/n_{ZnO:Al}$).

to the interface and studied the antireflecting behavior. We simulated $R_{tot}$ of this internal interface by applying ZnO:Al TCO as an incident medium and the p-µc-Si:H as the medium in transmission. The intensity of reflected light, which was calculated as a difference between the total intensity of light and the known intensity of applied incoming light, was monitored 300 nm away from the interface mean level. In Figure 6.3 (right) we show the results of $R_{tot}$ for three different 1-D gratings applied to the interface. The parameters of the gratings are triangular and rectangular in shape with period $P = 200$ nm and height $h = 70$ nm and triangular in shape with $P = 400$ nm and $h = 70$ nm. In addition, $R_{tot}$ of the flat interface is given for comparison. In the flat interface, the level of $R_{tot}$ is above 0.12 (12%). First, we analyze the effects of the triangular gratings. We can observe a significant reduction of $R_{tot}$ for the interfaces with the gratings, for both 400-nm and 200-nm periods in the entire wavelength region of interest. Further on we can see that the $R_{tot}$ corresponding to the triangular grating with the $P = 200$ nm is lower than the one with $P = 400$ nm. Taking into account these and other observations from the simulations with larger $P$ (up to 1 µm, not shown here), we can conclude that for enhancing the antireflecting effect, textures with small lateral dimensions ($P$) are important. Simulations indicated that the effect becomes especially pronounced if the $P < \lambda_0/n$ ($n$ – refractive index of the incident medium – ZnO:Al), that is, in the sub-wavelength region. Since the $n$ of the ZnO:Al is around 2, $P < \lambda_0/2$ in this case. Thus, for the $P = 200$ nm and $P = 400$ nm, the mentioned sub-wavelength condition starts to hold at $\lambda_0 < 400$ nm and $\lambda_0 < 800$ nm, respectively. The corresponding enhancements in the antireflecting effect (drops in the $R_{tot}$) are visible in Figure 6.3. The height of the grating feature was set to the value of 70 nm, which is close to the mentioned antireflecting condition $h = \lambda_0/4n$ (see Section 5.4.1, Chapter 5) for $\lambda_0 = 550$ nm.

Now we include also a simulation corresponding to the rectangular grating in the analysis. In Figure 6.3 we can observe that simulated $R_{tot}$ shows an even more pronounced antireflecting effect ($R_{tot} < 0.04$) than in the case of the triangular gratings. Rectangular grating seems to enable a better fulfillment of the antireflecting condition at $h = \lambda_0/4n$ for cancellation of reflected waves due to the horizontal orientation of top and bottom segments of texturization.

Next, we investigated the effect of light scattering at periodic textures. As a representative example we chose the case of light scattering at the back metal (Ag) reflector in a μc-Si:H solar cell. We studied a simplified structure of the back side of the cell that consists of an intrinsic μc-Si:H absorber layer directly stacked to the Ag back reflector with the periodic texture. With this structure we focus on the light-scattering effect only and eliminate back reflectances at front interfaces of the cell on purpose. As a measure of the level of light scattering at the μc-Si:H/Ag interface, the enhancement of light absorptance in a 500-nm thick zone of the μc-Si:H layer above the interface (Figures 6.4a and 6.4c) was determined (Figures 6.4b and 6.4d) and compared to the one obtained for a flat interface. Such indication of the level of light scattering inside the cell includes also the effects of near-field absorption, which has to be considered, since the region of observation is in light wavelength range. The simulated absorptance in μc-Si:H is thus attributed to (i) one pass of the perpendicularly applied incoming ("inc") ray propagating toward the interface (not dependent on the interface type) and (ii) one pass of the reflected scattered rays ("refl") that include information on scattering intensity and angles. The absorptance enhancement factor, $AEF$, due to light scattering at the interface, we define as given in Equation (6.2).

$$AEF = \frac{A_{\text{inc+refl GRATING}} - A_{\text{inc+refl FLAT}}}{A_{\text{inc+refl FLAT}}} \cdot 100\% \qquad (6.2)$$

The results of the $AEF$ for selected triangular and rectangular gratings applied to the μc-Si:H/Ag interface are plotted in Figure 6.4. To evaluate the $AEF$ of the periodic textures, the $AEF$ of selected randomly textured interface is plotted (pyramid-like texture, $\sigma_{rms} = 50$ nm). The simulator Sun*Shine* was used for this simulation (applying a semi-coherent 1-D model). Results presented in Figure 6.4b show that among the analyzed periodic textures, the triangular texture with $P = 200$ nm and $h = 100$ nm exhibits the highest $AEF$ (up to $\lambda_0 = 900$ nm), which is comparable to the $AEF$ of the randomly textured interface. However, we have to point out a distinct enhancement peak related to all periodic textures in the wavelength region of 650 nm to 750 nm.

This peak is much higher (up to 80% enhancement) than the level corresponding to the randomly textured interface and can be related to the appearance of the first diffraction order (for the grating with $P = 200$ nm) and the second diffraction order (for $P = 400$ nm) in μc-Si:H material ($n \approx 3.8$). In the

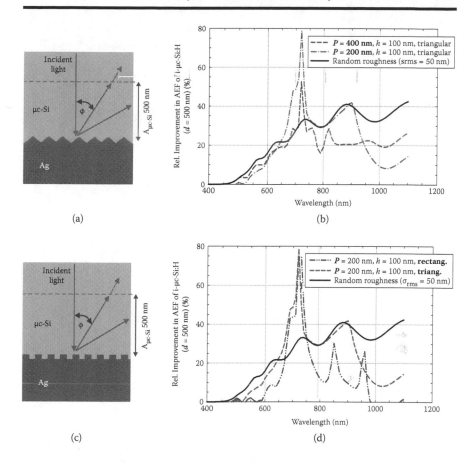

**FIGURE 6.4**    Simulated structures with (a) triangular and (c) rectangular grating. The corresponding simulated absorptance enhancement factors in a 500-nm thick zone of μc-Si:H absorber above the Ag textured reflector are shown in (b) and (d). The enhanced absorptance is related to light scattering at the textured μc-Si:H/Ag interface.

wavelength region of the peak, scattering of discrete rays into a large scattering angle (almost parallel to the interface plane) occurs. Comparing the scattering level of triangular and rectangular gratings (Figure 6.4d), the triangular ones exhibit a higher level in a broader region of wavelengths. A similar trend was confirmed by simulations of triangular and rectangular gratings with other periods and heights ($P \leq 1$ μm, $h \leq 300$ nm).

Further, we investigate another issue that appears to be important. These are optical losses in the Ag reflector related to different textures. For the Ag layer, realistic optical properties were considered in simulations. The results for the gratings with $P = 200$ nm and $h = 100$ nm and for the randomly textured Ag reflector are shown in Figure 6.5. For the triangular grating we can observe the

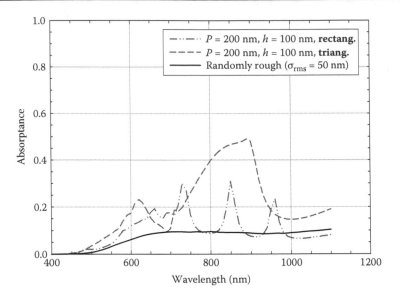

**FIGURE 6.5**   Simulated absorptances (optical losses) in Ag reflector with different textures.

highest absorptance in a broad wavelength range. On the other hand, the Ag reflector with triangular gratings showed high scattering abilities indicated by high *AEF*; however, its reflectivity is limited by parasitic absorption in Ag. The *AEF* implicitly includes reflectivity properties of the interface, showing over-all benefits of the triangular grating. Nevertheless, when introducing textures with intensified scattering properties to the metal films, one should be aware that optical losses may increase. One of the reasons for enhanced absorptance is excitation of surface plasmon polaritons (related to oscillations of electrons which movements are caused by the electromagnetic waves) (Atwater and Polman 2010). So-called parasitic plasmon absorption, decreasing the total reflectance of textured metal films (like Ag, Al, Au), presents a major drawback of metal reflectors in the optical sense. In a simulation of a randomly textured Ag reflector using the 1-D Sun*Shine* simulator (Figure 6.4d), the lowest absorptance is observed. However, this can be partially related to possible underestimation of the plasmonic effect in the 1-D simulation in this case, whereas in rigorous 2-D or 3-D simulations the plasmonic effects are included implicitly by considering realistic complex refractive indexes of metal layers and the geometry of the texture. In the semi-coherent 1-D simulation approach (such as the Sun*Shine* case) we can include them by reflectance correction function of the reflector (Krč and Topič 2011).

   Investigation of the presented optical effects (antireflection, scattering, parasitic absorption) helps us to understand and predict the optical situation in a complete solar cell structure.

## 6.2.3   SIMULATION OF A SOLAR CELL WITH PERIODICALLY TEXTURED INTERFACES

We carried out an optical simulation of a complete μc-Si:H solar cell in the following configuration (Figure 6.6a): ITO TCO (70 nm)/p-μc-Si:H (15 nm)/i-μc-Si:H (500 nm)/n-μc-Si:H (20 nm)/ZnO (100 nm)/Ag on textured substrate. The cell is in the substrate configuration and the periodic texture is introduced from the rear, substrate side. In these simulations of the μc-Si:H solar cell, we assume that the texture is ideally transferred from the substrate surface to all internal interfaces (only approximation of realistic case). Triangular grating was selected for the texture. Realistic optical properties of layers were used in 2-D simulations (Springer, Poruba, and Vanecek 2004). The simulated absorptances in the i-μc-Si:H absorber layers, $A_i$, are presented in Figure 6.6b for the triangular gratings with $P = 400$ nm and two different $h$ (300 nm and 100 nm). Comparing the simulation results for the cells with gratings to the cell with flat interfaces, noticeable enhancements in the $A_i$ are observed for the structures including gratings. At shorter wavelengths ($\lambda_0 < 600$ nm) the increase is related to the antireflecting effect at the front interfaces (air/ITO/p-μc-Si:H), whereas at longer wavelengths, both antireflecting and scattering effects determine the enhancement. In this case the scattering is present not only at the back interface, as analyzed in the previous section, but also at the front interfaces. In addition, light trapping due to reflectances of forward- and backward-going light at the interfaces is included in simulations of the complete solar cell structures.

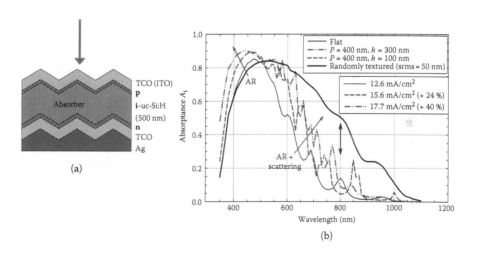

**FIGURE 6.6**   Simulated structure (a) and simulated absorptance (b) in the i-μc-Si:H absorber layer of complete thin μc-Si:H solar cells with the grating of triangular shape applied to all interfaces [see (a)]. Simulations of the cells with the gratings are compared to the cells with flat and randomly textured interfaces (root-mean-square roughness $\sigma_{rms} = 50$ nm).

We can observe noticeable enhancement in the $J_{SC}$ values for both cells with the gratings, compared to the flat structure (an increase of 24% to 40%, see inset in Figure 6.6).

In Figure 6.6 the calculated $A_i$ for the cell of the same structure but with randomly textured interfaces with $\sigma_{rms}$ = 50 nm (1-D simulation) is plotted in addition. Significantly higher $A_i$ is observed in this case for the structure with the randomly textured interfaces in the long-wavelength region. Again, the difference can be related partially to possible underestimation of plasmonic losses in the Ag reflector in the presented 1-D simulation, but as shown in the following, this is not the only reason. Additional simulations of the total reflectance $R_{tot}$ from the entire solar cell structure revealed that in the case of structures with the analyzed gratings, more long-wavelength light escapes from the structure than in the case of the cell with randomly textured interfaces. Thus, the trapping of light in the structures with the two gratings is less effective than in the case of the cell with randomly textured interfaces. Detailed optical analysis reveals that the main reason for this appears to be insufficient scattering (outside the specular direction) of the light beams which incidents the grating interface non-perpendicularly. The situation is illustrated in Figure 6.7. A simplified solar cell structure with flat front interfaces and without $p$- and $n$-doped layers is taken into consideration to demonstrate this effect more clearly. (Tests were done on a complete structure as well.)

First the light (beam 1) is efficiently scattered at the back periodically textured reflector (first incidence is perpendicular; according to Figure 6.4, the scattering on the grating can be comparable or even higher than in the case of the randomly textured interface). Then, selected scattered light beam (2) can be reflected back at the front interfaces (3). Further on, the beam approaches the textured back reflector again, but this time under a non-perpendicular incident angle. According to theory and simulations, this can result in reflected scattered beams propagating close to the specular direction (4) as in flat interfaces. The beams which fall onto the front interfaces under small incident angles (inside the escaping zone) cannot be efficiently trapped inside the structure.

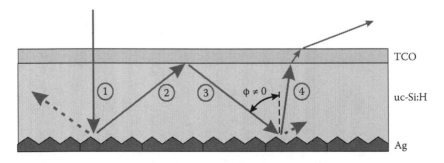

**FIGURE 6.7**   A schematic picture of light scattering at the back periodically textured reflector in a simplified solar cell structure.

Optical simulations enable tracing and investigating all of these effects individually and identifying the most critical issues as pointed out. From the presented results we can see that the optimization of periodic textures is required to ensure efficient light trapping in the cell that is comparable to or better than that in the cells with optimized random textures. First, different periods and heights as well as different shapes of the grating features have to be examined for different types of cells. To include all effects we have to carry out optical simulations and optimization on complete solar cell structures.

### 6.2.4  OPTIMIZATION OF ONE-DIMENSIONAL PERIODIC TEXTURES IN THIN-FILM SILICON SOLAR CELLS

Optimization of periodic textures is required in order to obtain maximal achievable gain in $J_{SC}$ in solar cells. In this section we apply 2-D optical simulation to optimize periodic textures for three different types of thin-film silicon solar cells: a-Si:H, µc-Si:H, and tandem micromorph a-Si:H/µc-Si:H. Let us carry out a systematic analysis in which four types of periodically textured profiles are included: (a) rectangular, (b) triangular, (c) sinusoidal, and (d) U-like (Figure 6.8). The period of the textures ($P$) and the vertical height ($h$) are chosen as the main optimization parameters. Additionally, the influence of the duty-cycle ($DC$), which is defined as the ratio of the positive slope to the entire period, is also investigated for the rectangular and triangular shape. Based on the simulation results, the optimal periodic textures with respect to the highest improvements in the $J_{SC}$ and $QE$ are determined for all three types of thin-film silicon solar cells.

First, we perform the optimization of surface texture in the a-Si:H solar cell, which has the following configuration (Figure 6.9): encapsulating EVA foil (incident medium)/ITO TCO (60 nm)/p-a-SiC:H (15 nm)/i-a-Si:H (200 nm)/n-a-Si:H (20 nm)/sputtered ZnO (40 nm)/Ag on textured substrate (either plastic or steel foil covered with embossed texture). The EVA foil was chosen as the incident medium to approach the optical situation on the PV module level. Due to

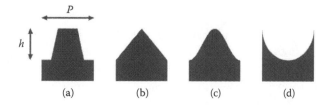

**FIGURE 6.8**  The analyzed periodic textures: (a) rectangular, (b) triangular, (c) sinusoidal, and (d) U-like. The period ($P$) and the height ($h$) parameters are also indicated. In the case of the rectangular shape we consider 80 degrees as the inclination angle of the steep walls.

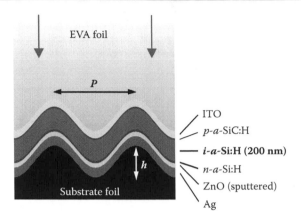

**FIGURE 6.9**   The analyzed a-Si:H solar cell structure on the substrate foil with sinusoidal (sine) texture.

relatively thin layers in the solar cell, the initial periodic texture of the substrate is considered to be translated ideally to all interfaces in the case of the a-Si:H cell. (The same texture was applied to all internal interfaces.) More details can be found in Lipovšek (2012).

In simulations we varied $P$ in the range of 200 nm to 1000 nm (step 100 nm), $h$ (150 nm to 600 nm, step 150 nm) and $DC$ (10%, 25%, 50%, 75%, 90%, and 100% for the rectangular and triangular texture).

In Figure 6.10 the contour plots of potential $J_{SC}$ increases related to the different textures and parameters $P$ and $h$ are shown. As a reference a solar cell with ideally flat interfaces rendering $J_{SC} = 11.1$, mA/cm² was considered. The increase in $J_{SC}$ was calculated according to Equation (6.3).

$$\Delta J_{SC} = \frac{J_{SC\,TEXT} - J_{SC\,FLAT}}{J_{SC\,FLAT}} \cdot 100\% \qquad (6.3)$$

A summary of the optimal combinations of $P$, $h$, and $DC$ from the results presented in Figure 6.10 and other simulations is given in the bar plot in Figure 6.11a. The highest improvements in $J_{SC}$ obtained for a-Si:H cell (> 30%) are observed for the triangular grating with $P = 200$ nm, $h = 300$ nm, $DC = 90\%$ and for the grating with sinusoidal shape with $P = 300$ nm and $h = 300$ nm. For rectangular grating, the optimal $DC$ was 50%. The triangular grating with $DC = 90\%$ (blazed profile) was found to minimize reflectance in specular direction for the non-perpendicular incident rays (effect 4 in Figure 6.7) due to a reduction of the negative diffraction order (Morf 1995). However, since fabrication of the grating with the blazed profile is more difficult than, for example, sinusoidal grating, where laser holographic methods can be applied on a large area (Rampal and Rampal 1993; K. Söderström et al. 2011), the latter is chosen as the

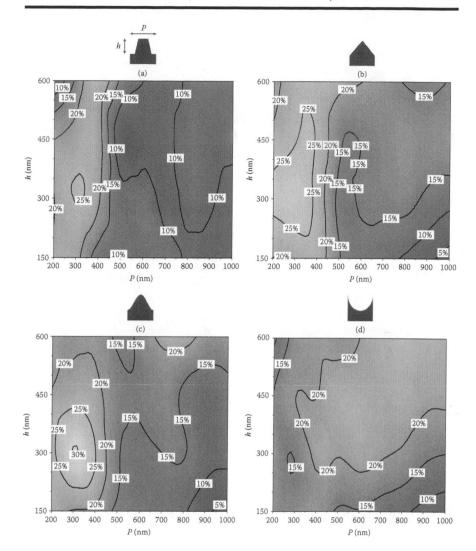

**FIGURE 6.10**    Simulated relative increase in short-circuit current density ($J_{SC}$) as a function of texturization shape, the period ($P$), and the height ($h$) of the texturization feature for the case of amorphous silicon solar cell structure. The reference presents an ideally flat cell with $J_{SC} = 11.1$ mA/cm$^2$.

favorite shape of basic periodic textures for a-Si:H cells. Such rounded textures are also favorable from the semiconductor layer growth point of view (Dubail et al. 2000).

For the a-Si:H solar cells with triangular (blazed), sinusoidal, and rectangular profiles with optimal $P$ and $h$, the $QE$ curves are shown in Figure 6.11b. As

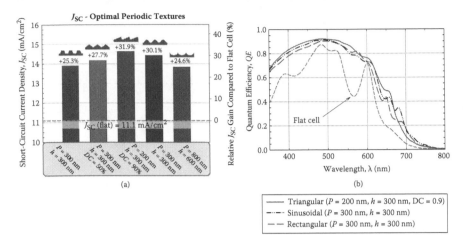

**FIGURE 6.11**   Selection of the optimal periodic textures and the corresponding $J_{SC}$ increases (a) and *QE*s of a-Si:H solar cells with different optimal ("best") periodic textures (b).

explained in the previous section, the effect of antireflection at front interfaces and light scattering at textured interfaces lead to enhanced *QE* in short- and long-wavelength parts of the spectrum.

As the second type of solar cell, a μc-Si:H cell in the following configuration is analyzed (Figure 6.12): EVA/LP-CVD ZnO:B (2 μm)/p-μc-Si:H (20 nm)/i-μc-Si:H (1.2 μm)/n-μc-Si:H (20 nm)/sputtered ZnO (60 nm)/Ag (100 nm)/textured substrate foil. The LP-CVD ZnO:B front contact presents an alternative to ITO TCO (T. Söderström et al. 2010).

In the case of μc-Si:H cell simulations, however, we did not consider the ideal transfer of the texture of the substrate to the front interfaces of the cell. The reason is crystalline growth of microcrystalline silicon layers, which smoothes and changes the initial roughness (Vallat-Sauvain et al. 2005). Further on, the growth of LP-CVD ZnO:B TCO forces pyramid-like texture at the front EVA/ZnO:B interface. Based on the models describing microcrystalline silicon growth (Vallat-Sauvain et al. 2005), we determined the surface morphology of the top surface of the 1.2-μm thick i-μc-Si:H layer, whereas the change in the morphology due to thin doped layers was neglected. Considering this, the sinusoidal texture of the substrate is, for example, transformed to a wing-like shape on the front. The change from the original texture is especially pronounced for low-*P*/high-*h* values.

To determine the texture of the front surface of the ZnO:B TCO, we considered naturally grown pyramidal (random) roughness of this material, which we approximated by periodic triangular grating, having average lateral and vertical dimensions of the texturization features as reported for the ZnO:B layer of such thickness (*P* = 200 nm, *h* = 240 nm) (Haug et al. 2010). Thus, the front interfaces of the μc-Si:H solar cell (EVA/ZnO:B/p-μc-Si:H/i-μc-Si:H) are determined

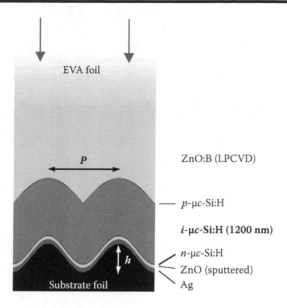

**FIGURE 6.12** The μc-Si:H solar cell structure on the substrate foil with sine texture.

mostly by the layer growth mechanisms, which, however, also depend on the substrate texture. To determine the optimal periodic texture of the substrate for the case of μc-Si:H solar cell structure, we varied $P$ (200–2000 nm), $h$ (300–1050 nm), and $DC$ (10%, 25%, 50%, 75%, 90%, 100%).

The $J_{SC}$ contour plots are shown in Figure 6.13. As a reference we considered the μc-Si:H solar cell with ideally flat interfaces, rendering $J_{SC}$ = 17.1 mA/cm². 

The increases in $J_{SC}$ corresponding to the optimal $P$ and $h$ combinations of particular grating shapes are summarized in a bar plot in Figure 6.14a.

The results show that the triangular texture with $P$ = 1000 nm, $h$ = 900 nm, and $DC$ = 0.5 in this case is most advantageous, achieving 36.0% higher $J_{SC}$ than the cell deposited on a flat substrate. Besides the cell with optimal triangular texture, the cells with optimized sinusoidal and blazed triangular texture exhibit only a bit lower enhancements in $J_{SC}$, whereas for U-like and especially rectangular, texture differences become larger in this case. Compared to the a-Si:H solar cell, we can observe larger values for optimal $P$ and $h$ for the μc-Si:H cells. This is because in μc-Si:H material, longer wavelengths of light have to be scattered ($\lambda_0$ > 700 nm) and absorbed, which can be achieved by the texturization features of larger dimensions.

The $QE$ curves determined for the μc-Si:H cells with the optimal triangular, sinusoidal, and rectangular textures (Figure 6.14b) indicate similar trends as those observed for the a-Si:H cells, most notably the beneficial combination of antireflective behavior in the short-wavelength region and efficient light scattering at longer wavelengths.

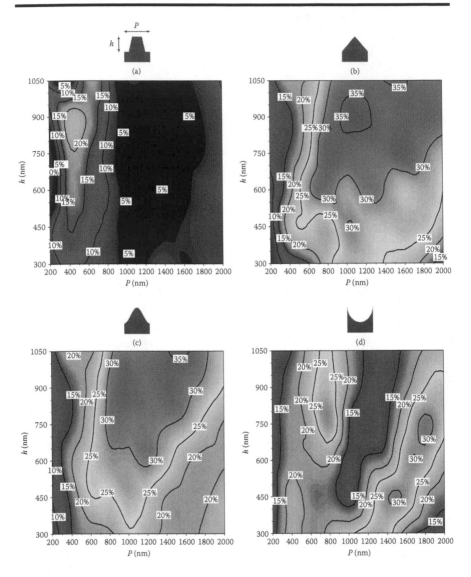

**FIGURE 6.13** Simulated relative increase in short-circuit current density ($J_{SC}$) as a function of texturization shape, the period ($P$), and the height ($h$) of the texturization feature for the case of the microcrystalline silicon solar cell structure. The reference presents an ideally flat cell with $J_{SC} = 17.1$ mA/cm$^2$.

As the third example we show the simulation results of optimization of the periodic texture for a micromorph solar cell in the following configuration (Figure 6.15): EVA /LP-CVD ZnO:B (2 μm)/p-a-SiC:H (15 nm)/i-a-Si:H (200 nm)/n-a-Si:H (20 nm)/p-μc-Si:H (20 nm)/i-μc-Si:H (1.2 μm)/n-μc-Si:H (20 nm)/ sputtered ZnO (60 nm)/Ag (100 nm)/textured substrate.

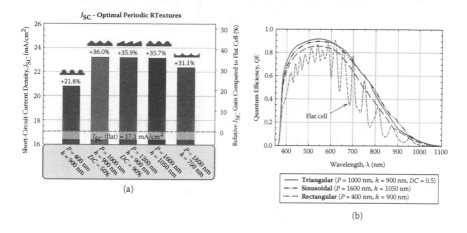

(a)

(b)

**FIGURE 6.14**   Selection of the optimal periodic textures with the corresponding $J_{SC}$ increases (a) and selected $QE$ curves (b) for the analyzed µc-Si:H cell.

As before, the change in the initial periodic texture has to be considered at the front surface of the i-µc-Si:H layer and other remaining front interfaces due to the crystalline growth of µc-Si:H and ZnO:B layers. To determine the improvements in $J_{SC}$ related to different surface textures, we varied the parameters of the texture as in the case of the µc-Si:H cell.

We monitored the $J_{SC}$ of the top and bottom cells separately. The current densities of the reference solar cell with ideally flat interfaces were $J_{SC\_top} = 10.37$ mA/cm² and $J_{SC\_bottom} = 8.03$ mA/cm² (not current matched). In Figure 6.16 the

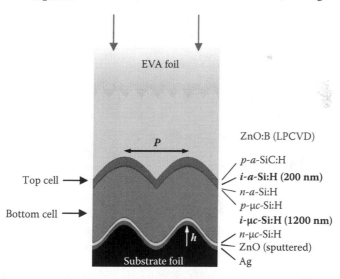

**FIGURE 6.15**   The a-Si:H/µc-Si:H solar cell structure deposited on substrate with sine texture.

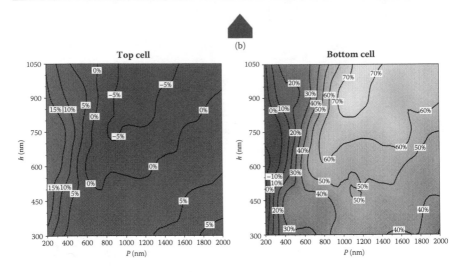

**FIGURE 6.16**  Simulated relative increase in short-circuit current density ($J_{SC}$) of the top (left) and the bottom cell (right) as a function of the period ($P$) and the height ($h$) for the case of the triangular texturization (b). The reference presents an ideally flat cell with $J_{SC\_top} = 10.37$ mA/cm$^2$ and $J_{SC\_bottom} = 8.03$ mA/cm$^2$.

contour plots showing the increases in $J_{SC\_top}$ and $J_{SC\_bottom}$ are presented for the case of triangular texture as a function of $P$ and $h$.

The summary of the optimal combinations of parameters and the corresponding increases in $J_{SC}$ are shown in the bar plot in Figure 6.17a. Simulations show that mostly $J_{SC\_bottom}$ is affected by the texturization of the substrate. A slight increase in $J_{SC\_top}$ can be observed if small periods (~300 nm) are introduced. The optimal parameters for achieving the highest increase in $J_{SC\_bottom}$ are the same as in the case of single-junction μc-Si:H cell (the highest increase for the triangular texture with $P = 1000$ nm, $h = 900$ nm, and $DC = 50\%$). However, the value of relative increase in $J_{SC\_bottom}$ is with respect to the reference flat cell ($J_{SC\_top} = 10.37$ mA/cm$^2$ and $J_{SC\_bottom} = 8.03$ mA/cm$^2$) and is much higher (74%) than in the case of single-junction μc-Si:H cell (36%). This is because in the micromorph cell, the μc-Si:H cell component is exposed mostly to the long-wavelength part of the spectrum, where light scattering has a high influence on the $J_{SC}$. The bar plot in Figure 6.17a and the $QE$ curves in Figure 6.17b show that the rectangular texture is less promising than the others.

As we know, in tandem solar cells the component that renders the lowest current determines the $J_{SC}$ of the entire device. In the analyzed case the lowest current has the bottom cell; thus, improvements in $J_{SC-bottom}$ mean also improvements in $J_{SC}$ of the cell.

Presented study of periodic surface textures in the three types of silicon solar cells and indicated increases in $J_{SC}$ and $QE$ are related to 1-D (linear) periodic textures. As shown in Section 4.2, Chapter 4, 2-D periodic textures, which

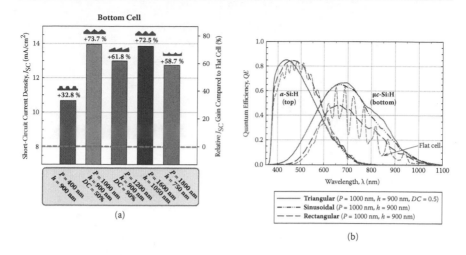

**FIGURE 6.17**   Selection of the optimal periodic textures with the corresponding $J_{SC}$ increases of the bottom cell (a) and $QE$ curves of top and bottom cells (b).

can be simulated by 3-D models, exhibit potential for larger improvements. However, similar indications for optimal dimensions ($P$, $h$) were identified also for 2-D textures; thus, the performed analysis of 1-D textures with 2-D simulations gives useful directions for selection of optimal geometry of the texture for specific types of solar cells.

Recent research results on 2-D periodic textures show that optimal periodic textures can even beat current state-of-the-art random textures (Isabella et al. 2008; Battaglia et al. 2012). By resolving problems explained in Section 6.2.3 improvements can be even greater. On the other hand, findings for optimal lateral and vertical parameters of periodic textures can indicate directions for improving random textures as well. Further work has to be done to develop reliable models of layer growth, which would enable us to describe the texture at each interface more accurately and also predict possible cracks in semiconductor layers which are intensified if high aspect ratio textures are used. The cracks can deteriorate electrical properties of devices significantly. Examples of models describing realistic layer growth for 2-D and 3-D optical simulations are already underway (Sever et al. 2012).

## 6.3   COMBINATION OF ONE-DIMENSIONAL PHOTONIC CRYSTAL STRUCTURES WITH ONE-DIMENSIONAL PERIODIC TEXTURE

As the last example of application of 2-D simulations we investigate a structure where we combine vertical periodicity, represented by 1-D photonic crystal (PC), and lateral periodicity, represented by 1-D diffraction gratings in one

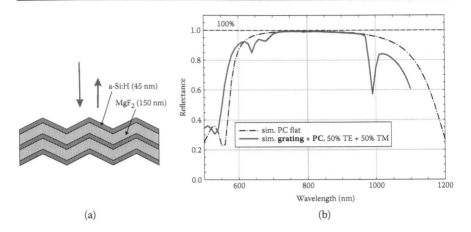

(a)                                                    (b)

**FIGURE 6.18**   (a) 1-D PC structure with periodically textured interfaces (triangular grating) and (b) simulated total reflectance of the PC with textured and flat interfaces.

structure. We get a kind of PC with periodically textured interfaces. In the next investigation we check whether such PC still renders high reflectivity properties, related to multiple reflectance/transmittance processes at internal interfaces. We investigated such structures since the texturization of interfaces is usually present in thin-film solar cells and therefore transferred also to PC structures (either intermediate or back reflector). As a test case we take a simple PC structure as shown in Figure 6.18a. It consists of three high-refractive index material (i-a-Si:H) and two low-refractive index material (MgF$_2$), presenting two-and-a-half period 1-D PC. As the lateral periodicity we introduce a triangular grating ($P = 700$ nm, $h = 100$ nm).

Simulation results of total reflectance of the structure for perpendicular incidence of light are shown in Figure 6.18b. $R_{tot}$ of the textured structure is compared to the $R_{tot}$ of a flat PC. The flat PC structure enables us to achieve high reflectance close to 100% in the wavelength region of 700 nm to 900 nm as shown by simulations. We can observe that the introduced texture does not decrease the level of reflectance behavior; however, the changes are observed and can be characterized by (i) the shift of the region of high reflectance toward shorter wavelengths and (ii) sharp drops in reflectance characteristics. We can relate effect 1 to the increase of optical thicknesses of layers due to inclination of the interfaces (actual thickness is defined perpendicularly to the inclined interface segments, whereas the optical thickness is defined according to the direction of light propagation). Effect 2 is related to the appearance of discrete scattering orders due to introduced grating inside the structure, leading to enhanced optical losses in the structure. Thus, such structures have to be carefully designed to benefit from both effects, high reflectance and introduced scattering to the solar cell structures. As in other cases, optical modeling and simulations can be employed in optimization. Experimentally, periodic textures in combination with PC were studied in Isabella et al. (2009).

In this chapter we presented selected highlights of possible use of 2-D optical simulations. The spectrum of possible applications is broad, so selected issues can be the subject of the reader's own investigations and future trends in thin-film photovoltaics.

## REFERENCES

Atwater, H. A., and A. Polman. 2010. "Plasmonics for Improved Photovoltaic Devices." *Nature Materials* 9 (3): 205–213. doi:10.1038/nmat2629.

Battaglia, C., C.-M. Hsu, K. Soederstroem, J. Escarre, F.-J. Haug, M. Charriere, M. Boccard, et al. 2012. "Light Trapping in Solar Cells: Can Periodic Beat Random?" *Acs Nano* 6 (3) (March): 2790–2797. doi:10.1021/nn300287j.

Battaglia, C., J. Escarré, K. Söderström, M. Charrière, M. Despeisse, F.-J. Haug, and C. Ballif. 2011. "Nanomoulding of Transparent Zinc Oxide Electrodes for Efficient Light Trapping in Solar Cells." *Nature Photonics* 5 (9) (September): 535–538. doi:10.1038/NPHOTON.2011.198.

Beckmann, P., and A. Spizzichino. 1987. *The Scattering of Electromagnetic Waves from Rough Surfaces.* Artech Print on Demand.

Čampa, A. 2009. *Modelling and Optimisation of Advanced Optical Concepts in Thin-Film Solar Cells.* PhD thesis, University of Ljubljana.

Dewan, R., I. Vasilev, V. Jovanov, and D. Knipp. 2011. "Optical Enhancement and Losses of Pyramid Textured Thin-Film Silicon Solar Cells." *Journal of Applied Physics* 110 (1) (July 1). doi:10.1063/1.3602092.

Dubail, J., E. Vallat-Sauvain, J. Meier, S. Dubail, and A. Shah. 2000. "Microstructures of Microcrystalline Silicon Solar Cells Prepared by Very High Frequency Glow-Discharge." *MRS Online Proceedings Library* 609: 609–A13.6. doi:10.1557/PROC-609-A13.6.

Haase, C., and H. Stiebig. 2006. "Optical Properties of Thin-Film Silicon Solar Cells with Grating Couplers." *Progress in Photovoltaics: Research and Applications* 14 (7): 629–641. doi:10.1002/pip.694.

Haase, C., and H. Stiebig. 2007. "Thin-Film Silicon Solar Cells with Efficient Periodic Light Trapping Texture." *Applied Physics Letters* 91: 061116. doi:10.1063/1.2768882.

Haug, F., C. Battaglia, D. Dominé, and C. Ballif. 2010. "Light Scattering at Nano-textured Surfaces in Thin Film Silicon Solar Cells." In *Proceedings of the Thirty-Fifth IEEE Photovoltaic Specialists Conference (PVSC),* 2010, 000754–00759. doi:10.1109/PVSC.2010.5617091.

Isabella, O., A. Čampa, M. C. R. Heijna, W. Soppe, A. J. M. van Erven, R. H. Franken, H. Borg, M. Zeman, 2008. "Diffraction Gratings for Light Trapping in Thin-Film Silicon Solar Cells." 23rd European Photovoltaic Solar Energy Conference and Exhibition, Valencia, Spain. WIP Renewable Energies (DOI: 10.4229/23rdEUPVSEC20083AV.1.48).

Isabella, O., J. Krc, M. Zeman. 2009. "Application of Photonic Crystals as Back Reflectors in Thin-Film Silicon Solar Cells." 24th European Photovoltaic Solar Energy Conference, September 21–25, 2009, Hamburg, Germany, WIP Renewable Energies (DOI: 10.4229/24thEUPVSEC2009-3BO.9.5).

Krč, J., and M. Topič. 2011. "Sun*Shine* Optical Simulator V1.2.8—User's Manual."

Lipovšek B. 2012. "Advanced light management in thin-film optoelectronic devices." PhD Thesis, University of Ljubljana, Slovenia.

Lipovšek, B., M. Cvek, A. Čampa, J. Krč, and M. Topič. 2010. "Analysis and Optimisation of Periodic Interface Textures in Thin-Film Silicon Solar Cells." 25th European Photovoltaic Solar Energy Conference and Exhibition, September 6–10, 2010, Valencia, Spain. WIP Renewable Energies, pp. 3120–3123.

Morf, R. H. 1995. "Exponentially Convergent and Numerically Efficient Solution of Maxwell's Equations for Lamellar Gratings." *Journal of the Optical Society of America A* 12 (5): 1043–1056. doi:10.1364/JOSAA.12.001043.

Rampal, V. V., and V. V. Rampal. 1993. *Lasers & Holography.* World Scientific Publishing.

Sever, M., B. Lipovšek, J. Krč, M. Topič. 2012. "Optimisation of Surface Textures in Thin-Film Silicon Solar Cells with 3D Optical Modelling by Considering Realistic Layer Growth." 27th European Photovoltaic Solar Energy Conference and Exhibition, Frankfurt, Germany, September 24–28, 2012. WIP Renewable Energies, pp. 2129-2131.

Söderström, K., J. Escarré, O. Cubero, F.-J. Haug, S. Perregaux, and C. Ballif. 2011. "UV-nano-imprint Lithography Technique for the Replication of Back Reflectors for *N-i-p* Thin Film Silicon Solar Cells." *Progress in Photovoltaics: Research and Applications* 19 (2): 202–210. doi:10.1002/pip.1003.

Söderström, T., F.-J. Haug, V. Terrazzoni-Daudrix, and C. Ballif. 2010. "Flexible Micromorph Tandem *a*-Si/*μc*-Si Solar Cells." *Journal of Applied Physics* 107 (1): 014507. doi:10.1063/1.3275860.

Soppe, W. J., H. Borg, B. B. Van Aken, C. Devilee, M. Dörenkämper, M. Goris, M. C. R. Heijna, J. Löffler, and P. Peeters. 2011. "Roll to Roll Fabrication of Thin Film Silicon Solar Cells on Nano-Textured Substrates." *Journal of Nanoscience and Nanotechnology* 11 (12) (December): 10604–10609. doi:10.1166/jnn.2011.4075.

Springer, J., A. Poruba, and M. Vaneček. 2004. "Improved Three-Dimensional Optical Model for Thin-Film Silicon Solar Cells." *Journal of Applied Physics* 96 (9) (November 1): 5329–5337. doi:10.1063/1.1784555.

Vallat-Sauvain, E., J. Bailat, J. Meier, X. Niquille, U. Kroll, and A. Shah. 2005. "Influence of the Substrate's Surface Morphology and Chemical Nature on the Nucleation and Growth of Microcrystalline Silicon." *Thin Solid Films* 485 (1–2): 77–81. doi:10.1016/j.tsf.2005.03.017.

# Index

**Note:** Page numbers ending in "e" refer to equations. Page numbers ending in "f" refer to figures. Page numbers ending in "t" refer to tables.